Pitman Research Notes in Mathematics Series

Submission of proposals for consideration
Suggestions for publication, in the form of outlines and representative samples, are invited by the Editorial Board for assessment. Intending authors should approach one of the main editors or another member of the Editorial Board, citing the relevant AMS subject classifications. Alternatively, outlines may be sent directly to the publisher's offices. Refereeing is by members of the board and other mathematical authorities in the topic concerned, throughout the world.

Preparation of accepted manuscripts
On acceptance of a proposal, the publisher will supply full instructions for the preparation of manuscripts in a form suitable for direct photo-lithographic reproduction. Specially printed grid sheets are provided and a contribution is offered by the publisher towards the cost of typing. Word processor output, subject to the publisher's approval, is also acceptable.

Illustrations should be prepared by the authors, ready for direct reproduction without further improvement. The use of hand-drawn symbols should be avoided wherever possible, in order to maintain maximum clarity of the text.

The publisher will be pleased to give any guidance necessary during the preparation of a typescript, and will be happy to answer any queries.

Important note
In order to avoid later retyping, intending authors are strongly urged not to begin final preparation of a typescript before receiving the publisher's guidelines and special paper. In this way it is hoped to preserve the uniform appearance of the series.

Longman Scientific & Technical
Longman House
Burnt Mill
Harlow, Essex, UK
(tel (0279) 26721)

Titles in this series

52 Wave propagation in viscoelastic media
F Mainardi

53 Nonlinear partial differential equations and their applications: Collège de France Seminar. Volume I
H Brezis and J L Lions

54 Geometry of Coxeter groups
H Hiller

55 Cusps of Gauss mappings
T Banchoff, T Gaffney and C McCrory

56 An approach to algebraic K-theory
A J Berrick

57 Convex analysis and optimization
J-P Aubin and R B Vintner

58 Convex analysis with applications in the differentiation of convex functions
J R Giles

59 Weak and variational methods for moving boundary problems
C M Elliott and J R Ockendon

60 Nonlinear partial differential equations and their applications: Collège de France Seminar. Volume II
H Brezis and J L Lions

61 Singular systems of differential equations II
S L Campbell

62 Rates of convergence in the central limit theorem
Peter Hall

63 Solution of differential equations by means of one-parameter groups
J M Hill

64 Hankel operators on Hilbert space
S C Power

65 Schrödinger-type operators with continuous spectra
M S P Eastham and H Kalf

66 Recent applications of generalized inverses
S L Campbell

67 Riesz and Fredholm theory in Banach algebra
B A Barnes, G J Murphy, M R F Smyth and T T West

68 Evolution equations and their applications
F Kappel and W Schappacher

69 Generalized solutions of Hamilton-Jacobi equations
P L Lions

70 Nonlinear partial differential equations and their applications: Collège de France Seminar. Volume III
H Brezis and J L Lions

71 Spectral theory and wave operators for the Schrödinger equation
A M Berthier

72 Approximation of Hilbert space operators I
D A Herrero

73 Vector valued Nevanlinna Theory
H J W Ziegler

74 Instability, nonexistence and weighted energy methods in fluid dynamics and related theories
B Straughan

75 Local bifurcation and symmetry
A Vanderbauwhede

76 Clifford analysis
F Brackx, R Delanghe and F Sommen

77 Nonlinear equivalence, reduction of PDEs to ODEs and fast convergent numerical methods
E E Rosinger

78 Free boundary problems, theory and applications. Volume I
A Fasano and M Primicerio

79 Free boundary problems, theory and applications. Volume II
A Fasano and M Primicerio

80 Symplectic geometry
A Crumeyrolle and J Grifone

81 An algorithmic analysis of a communication model with retransmission of flawed messages
D M Lucantoni

82 Geometric games and their applications
W H Ruckle

83 Additive groups of rings
S Feigelstock

84 Nonlinear partial differential equations and their applications: Collège de France Seminar. Volume IV
H Brezis and J L Lions

85 Multiplicative functionals on topological algebras
T Husain

86 Hamilton-Jacobi equations in Hilbert spaces
V Barbu and G Da Prato

87 Harmonic maps with symmetry, harmonic morphisms and deformations of metrics
P Baird

88 Similarity solutions of nonlinear partial differential equations
L Dresner

89 Contributions to nonlinear partial differential equations
C Bardos, A Damlamian, J I Díaz and J Hernández

90 Banach and Hilbert spaces of vector-valued functions
J Burbea and P Masani

91 Control and observation of neutral systems
D Salamon

92 Banach bundles, Banach modules and automorphisms of C*-algebras
M J Dupré and R M Gillette

93 Nonlinear partial differential equations and their applications: Collège de France Seminar. Volume V
H Brezis and J L Lions

94 Computer algebra in applied mathematics: an introduction to MACSYMA
R H Rand

95 Advances in nonlinear waves. Volume I
L Debnath

96 FC-groups
M J Tomkinson

97 Topics in relaxation and ellipsoidal methods
M Akgül

98 Analogue of the group algebra for topological semigroups
H Dzinotyiweyi

99 Stochastic functional differential equations
S E A Mohammed

100 Optimal control of variational inequalities
V Barbu
101 Partial differential equations and dynamical systems
W E Fitzgibbon III
102 Approximation of Hilbert space operators. Volume II
C Apostol, L A Fialkow, D A Herrero and D Voiculescu
103 Nondiscrete induction and iterative processes
V Ptak and F-A Potra
104 Analytic functions – growth aspects
O P Juneja and G P Kapoor
105 Theory of Tikhonov regularization for Fredholm equations of the first kind
C W Groetsch
106 Nonlinear partial differential equations and free boundaries. Volume I
J I Díaz
107 Tight and taut immersions of manifolds
T E Cecil and P J Ryan
108 A layering method for viscous, incompressible L_p flows occupying R^n
A Douglis and E B Fabes
109 Nonlinear partial differential equations and their applications: Collège de France Seminar. Volume VI
H Brezis and J L Lions
110 Finite generalized quadrangles
S E Payne and J A Thas
111 Advances in nonlinear waves. Volume II
L Debnath
112 Topics in several complex variables
E Ramírez de Arellano and D Sundararaman
113 Differential equations, flow invariance and applications
N H Pavel
114 Geometrical combinatorics
F C Holroyd and R J Wilson
115 Generators of strongly continuous semigroups
J A van Casteren
116 Growth of algebras and Gelfand–Kirillov dimension
G R Krause and T H Lenagan
117 Theory of bases and cones
P K Kamthan and M Gupta
118 Linear groups and permutations
A R Camina and E A Whelan
119 General Wiener–Hopf factorization methods
F-O Speck
120 Free boundary problems: applications and theory, Volume III
A Bossavit, A Damlamian and M Fremond
121 Free boundary problems: applications and theory, Volume IV
A Bossavit, A Damlamian and M Fremond
122 Nonlinear partial differential equations and their applications: Collège de France Seminar. Volume VII
H Brezis and J L Lions
123 Geometric methods in operator algebras
H Araki and E G Effros
124 Infinite dimensional analysis–stochastic processes
S Albeverio
125 Ennio de Giorgi Colloquium
P Krée
126 Almost-periodic functions in abstract spaces
S Zaidman
127 Nonlinear variational problems
A Marino, L Modica, S Spagnolo and M Degiovanni
128 Second-order systems of partial differential equations in the plane
L K Hua, W Lin and C-Q Wu
129 Asymptotics of high-order ordinary differential equations
R B Paris and A D Wood
130 Stochastic differential equations
R Wu
131 Differential geometry
L A Cordero
132 Nonlinear differential equations
J K Hale and P Martinez-Amores
133 Approximation theory and applications
S P Singh
134 Near-rings and their links with groups
J D P Meldrum
135 Estimating eigenvalues with *a posteriori/a priori* inequalities
J R Kuttler and V G Sigillito
136 Regular semigroups as extensions
F J Pastijn and M Petrich
137 Representations of rank one Lie groups
D H Collingwood
138 Fractional calculus
G F Roach and A C McBride
139 Hamilton's principle in continuum mechanics
A Bedford
140 Numerical analysis
D F Griffiths and G A Watson
141 Semigroups, theory and applications. Volume I
H Brezis, M G Crandall and F Kappel
142 Distribution theorems of L-functions
D Joyner
143 Recent developments in structured continua
D De Kee and P Kaloni
144 Functional analysis and two-point differential operators
J Locker
145 Numerical methods for partial differential equations
S I Hariharan and T H Moulden
146 Completely bounded maps and dilations
V I Paulsen
147 Harmonic analysis on the Heisenberg nilpotent Lie group
W Schempp
148 Contributions to modern calculus of variations
L Cesari
149 Nonlinear parabolic equations: qualitative properties of solutions
L Boccardo and A Tesei
150 From local times to global geometry, control and physics
K D Elworthy

151 A stochastic maximum principle for optimal control of diffusions
U G Haussmann
152 Semigroups, theory and applications. Volume II
H Brezis, M G Crandall and F Kappel
153 A general theory of integration in function spaces
P Muldowney
154 Oakland Conference on partial differential equations and applied mathematics
L R Bragg and J W Dettman
155 Contributions to nonlinear partial differential equations. Volume II
J I Díaz and P L Lions
156 Semigroups of linear operators: an introduction
A C McBride
157 Ordinary and partial differential equations
B D Sleeman and R J Jarvis
158 Hyperbolic equations
F Colombini and M K V Murthy
159 Linear topologies on a ring: an overview
J S Golan
160 Dynamical systems and bifurcation theory
M I Camacho, M J Pacifico and F Takens
161 Branched coverings and algebraic functions
M Namba
162 Perturbation bounds for matrix eigenvalues
R Bhatia
163 Defect minimization in operator equations: theory and applications
R Reemtsen
164 Multidimensional Brownian excursions and potential theory
K Burdzy
165 Viscosity solutions and optimal control
R J Elliott
166 Nonlinear partial differential equations and their applications. Collège de France Seminar. Volume VIII
H Brezis and J L Lions
167 Theory and applications of inverse problems
H Haario
168 Energy stability and convection
G P Galdi and B Straughan
169 Additive groups of rings. Volume II
S Feigelstock
170 Numerical analysis 1987
D F Griffiths and G A Watson
171 Surveys of some recent results in operator theory. Volume I
J B Conway and B B Morrel
172 Amenable Banach algebras
J-P Pier
173 Pseudo-orbits of contact forms
A Bahri
174 Poisson algebras and Poisson manifolds
K H Bhaskara and K Viswanath
175 Maximum principles and eigenvalue problems in partial differential equations
P W Schaefer
176 Mathematical analysis of nonlinear, dynamic processes
K U Grusa
177 Cordes' two-parameter spectral representation theory
D F McGhee and R H Picard
178 Equivariant K-theory for proper actions
N C Phillips
179 Elliptic operators, topology and asymptotic methods
J Roe
180 Nonlinear evolution equations
J K Engelbrecht, V E Fridman and E N Pelinovski
181 Nonlinear partial differential equations and their applications. Collège de France Seminar. Volume IX
H Brezis and J L Lions
182 Critical points at infinity in some variational problems
A Bahri
183 Recent developments in hyperbolic equations
L Cattabriga, F Colombini, M K V Murthy and S Spagnolo
184 Optimization and identification of systems governed by evolution equations on Banach space
N U Ahmed
185 Free boundary problems: theory and applications. Volume I
K H Hoffmann and J Sprekels
186 Free boundary problems: theory and applications. Volume II
K H Hoffmann and J Sprekels
187 An introduction to intersection homology theory
F Kirwan
188 Derivatives, nuclei and dimensions on the frame of torsion theories
J S Golan and H Simmons
189 Theory of reproducing kernels and its applications
S Saitoh
190 Volterra integrodifferential equations in Banach spaces and applications
G Da Prato and M Iannelli
191 Nest algebras
K R Davidson
192 Surveys of some recent results in operator theory. Volume II
J B Conway and B B Morrel
193 Nonlinear variational problems. Volume II
A Marino and M K Murthy
194 Stochastic processes with multidimensional parameter
M E Dozzi
195 Prestressed bodies
D Iesan
196 Hilbert space approach to some classical transforms
R H Picard
197 Stochastic calculus in application
J R Norris
198 Radical theory
B J Gardner
199 The C* – algebras of a class of solvable Lie groups
X Wang

200 Stochastic analysis, path integration and dynamics
D Elworthy
201 Riemannian geometry and holonomy groups
S Salamon
202 Strong asymptotics for extremal errors and polynomials associated with Erdös type weights
D S Lubinsky
203 Optimal control of diffusion processes
V S Borkar
204 Rings, modules and radicals
B J Gardner
205 Numerical studies for nonlinear Schrödinger equations
B M Herbst and J A C Weideman
206 Distributions and analytic functions
R D Carmichael and D Mitrović
207 Semicontinuity, relaxation and integral representation in the calculus of variations
G Buttazzo
208 Recent advances in nonlinear elliptic and parabolic problems
P Bénilan, M Chipot, L Evans and M Pierre
209 Model completions, ring representations and the topology of the Pierce sheaf
A Carson

Giuseppe Buttazzo

University of Ferrara

Semicontinuity, relaxation and integral representation in the calculus of variations

Longman Scientific & Technical

Copublished in the United States with
John Wiley & Sons, Inc., New York

Longman Scientific & Technical,
Longman Group UK Limited,
Longman House, Burnt Mill, Harlow
Essex CM20 2JE, England
and Associated Companies throughout the world.

Copublished in the United States with
John Wiley & Sons, Inc., 605 Third Avenue, New York, NY 10158

First published 1989

AMS Subject Classification: (Main) 49 A 50, 49-02, 49 A 10
(Subsidiary) 49 A 22, 28 A 33, 47 H 99

ISSN 0269-3674

British Library Cataloguing in Publication Data
Buttazzo, Giuseppe
Semicontinuity, relaxation and integral representation in the calculus of variations
1. Calculus of variations
I. Title.
515'.64
ISBN 0-582-01859-5

Library of Congress Cataloging-in-Publication Data
Buttazzo, Giuseppe.
Semicontinuity, relaxation, and integral representation in the calculus of variations
p. c. -- (Pitman research notes in mathematics series, ISSN 0269-3674: 207)
Bibliography: p.
Includes index.
ISBN 0-470-21343-4
1. Calculus of variations. I. Title. II. Series
QA315.B87 1989
515'.64dc19 88-38153 CIP

Printed and bound in Great Britain
by Biddles Ltd, Guildford and King's Lynn

Contents

Preface and Acknowledgments

The story of this book started at the end of 1985, when I was invited at the CMAF of Lisbon to give a short course on the Calculus of Variations; on that occasion I wrote some notes which appeared in a preprint form on the series "Textos e Notas" by the CMAF. I am indebted to J.F.Rodrigues and to M.L.Mascareñas for the help they gave me during the preparation of those notes, and to Professor H.Brezis who, some months later, encouraged me to write a more extended version, and put me in touch with Longman and the Research Notes in Mathematics Series.

The book contains an overview of work in the last few years on the classic Direct Method of the Calculus of Variations and on the more recent Relaxation Method for treating not well posed problems; the development of the theory has been possible due to the suggestions and supervision of Professor E.De Giorgi to whom I address my best acknowledgments.

I wish to thank my colleagues of Scuola Normale of Pisa and of the University of Pisa for the friendly and productive atmosphere we have experienced for so many years and for the time they spent listening to my problems. In particular, I thank L. Ambrosio and G.Dal Maso, who collaborated with me in several papers and whose results are presented in this book.

I wish to also thank H.Attouch, B.Dacorogna and R.V.Kohn; the discussions I had with them allowed me to learn the different perspectives followed by various schools in other countries.

Finally, I am grateful to my son Dario for being very quiet during his first two years of life which helped in the writing of this book.

Giuseppe Buttazzo

Introduction

In these last years there has been a very big development of the Calculus of Variations, especially for what concerns the existence theory, the necessary conditions for optimality, and the regularity of minimizers. In the classical framework the functionals considered are of the form

$$F(u) = \int_\Omega f(x,u(x),Du(x))\,dx \tag{0.1}$$

where $u(x)$ is a real-valued or a vector-valued function, and the hypotheses on the integrand $f(x,s,z)$ may vary, depending on the situation.

The most classical way to obtain the existence of a minimizer is the so-called Direct Method, and consists in proving for the functional F the lower semicontinuity and the coerciveness with respect to a suitable topology. In the most part of cases deriving from the applications, the coerciveness property is very easy to prove, and follows from estimates from below on the functional F, like

$$F(u) \geq \alpha\int_\Omega \left[|Du|^p+|u|^p\right] dx \qquad (\alpha>0,\ p>1).$$

In this sense, we may say the coerciveness is a "quantitative" condition. On the contrary, the lower semicontinuity condition, involves "qualitative" properties of the integrand $f(x,s,z)$ (as convexity or quasi-convexity with respect to z), and in some situations may fail. Nevertheless, it may be interesting to study the behaviour of minimizing sequences by characterizing their limit points as minimizers of a new functional ΓF (called relaxed functional of F).

This is the basic idea of relaxation, which we shall apply to several situations in next chapters. The main difficulty of this approach consists in the fact that the relaxed functional ΓF is just defined in an abstract topological way (as the lower semicontinuous envelope of the functional F), whereas we would like to deal with a functional represented in some integral form. This goal will be achieved by suitable integral representation theorems which, in the case of functionals like (0.1), give

$$\Gamma F(u) = \int_\Omega \varphi_f(x,u(x),Du(x))\,dx$$

where φ_f is an integrand depending on f and which may be explicitely computed in many cases.

The main goal of this book is to study the lower semicontinuity, relaxation, and integral representation results for several kinds of problems in the Calculus of Variations.

The first chapter is devoted to topological preliminaries and to the abstract setting of the Direct Method and of the Relaxation Method in general topological spaces; we shall consider also the framework of Banach spaces endowed with sequential and topological weak and weak* convergences, and at the end of the chapter some classical examples are treated.

In Chapter 2 we consider functionals of the form

$$(0.2) \qquad F(u,v) = \int_\Omega f(x,u(x),v(x))\,d\mu(x)$$

where μ is a given measure, and u,v vary in suitable L^p spaces. The function v(x) is considered as independent of u(x) and, even if this is not the case of functionals like (0.1) where u and v are related by the equation v(x)=Du(x), the knowledge of lower semicontinuity and relaxation results for (0.2) may be important also for the study of (0.1). The relaxation result (with respect to the strong-weak convergence on pairs (u,v))

$$\Gamma F(u,v) = \int_\Omega \varphi(x,u(x),v(x))\, d\mu(x)$$

is obtained by proving that every lower semicontinuous mapping $F(u,v,B)$, defined for every (u,v) in some L^p spaces and for every μ-measurable set B, is actually an integral functional of the form (0.2) provided some additivity and locality properties are fulfilled.

Chapter 3 deals with functionals defined on the space $M(\Omega;\mathbf{R}^n)$ of bounded vector measures; starting from functionals of the form

$$F(\lambda) = \begin{cases} \displaystyle\int_\Omega f(x,u)\, d\mu & \text{if } \lambda=u\cdot\mu \text{ with } u\in L^1(\Omega;\mathbf{R}^n) \\ +\infty & \text{otherwise ,} \end{cases}$$

we shall prove that the relaxed functional ΓF (with respect to the weak* convergence in $M(\Omega;\mathbf{R}^n)$) can be written as an integral

$$\text{(0.3)} \qquad \Gamma F(\lambda) = \int_\Omega \varphi(x,\frac{d\lambda}{d\mu})\, d\mu + \int_\Omega \varphi^\infty(x,\frac{d\lambda^s}{d|\lambda^s|})\, d|\lambda^s|$$

for a suitable convex integrand φ depending on f, where $\varphi^\infty(x,\cdot)$ is the recession function of $\varphi(x,\cdot)$ and the measure $\lambda\in M(\Omega;\mathbf{R}^n)$ is split into the sum

$$\lambda = \frac{d\lambda}{d\mu}\cdot\mu + \lambda^s$$

by using the Lebesgue-Nikodym decomposition theorem. Again, this relaxation result is obtained via an integral representation theorem which proves that functionals of the form (0.3) are the only lower semicontinuous mappings $F(\lambda,B)$ (with $\lambda\in M(\Omega;\mathbf{R}^n)$ and B μ-measurable) having some appropriate additivity and locality properties.

Chapter 4 is devoted to classical functionals of the Calculus of Variations, that is functionals of the form (0.1) defined on some Sobolev spaces $W^{1,p}(\Omega;\mathbf{R}^m)$. In this case, the lower semicontinuity in the scalar case $m=1$ and in the vector-valued case

$m>1$ leads to two different conditions that are respectively the convexity and the so-called quasiconvexity of the integrand $f(x,s,z)$ with respect to z. As in the previous chapters the relaxed functional ΓF is written in the form

$$\Gamma F(u) = \int_\Omega \varphi(x,u(x),Du(x))\,dx$$

by using an integral representation theorem on Sobolev spaces. This chapter contains also a lower semicontinuity theorem which has been recently obtained for autonomous functionals like

$$\int_\Omega f(u,Du)\,dx$$

where the integrand $f(s,z)$ may have high discontinuities in the variable s.

The last chapter of the book (Chapter 5) is concerned with relaxation in Optimal Control Theory. Here the functionals we are dealing with are of the form

$$F(u,y) = \begin{cases} \displaystyle\int_\Omega f(x,y(x),u(x))\,dx & \text{if } y=L(u) \\ +\infty & \text{otherwise}, \end{cases}$$

where the control variable u belongs to some L^p spaces, the state variable y is in a Sobolev space, and the state equation $y = Lu$ is a differential equation which relates y to u. An interesting feature of this case is that the relaxed functional may have a differential constraint of the form $y \in \Lambda u$ where Λ is in general a multi-valued mapping.

This book does not deal with general lower semicontinuity, relaxation, and integral representation problems in the space of functions with bounded variation, even if some results in the one-dimensional case can be deduced from Chapter 3. The theory of functionals defined on the space $BV(\Omega;\mathbf{R}^m)$ is in rapid evolution, and it would require an entire book just in order to give the basic framework. On the other hand, the most stable part of the theory of BV functions is already written in some good books as for instance Federer [151], Giusti [171], Massari & Miranda [219].

Finally, I want to remark that while this book deals with problems involving a single functional, it is possible to develop a theory in which sequences of functionals are considered, and the asymptotic behaviour of minimum points (instead of the existence of a minimum point) is studied. This is the theory of Γ-convergence, which is involved in all cases in which one wants to deduce the asymptotic behaviour of the solutions of a sequence of variational problems from the asymptotic behaviour of the related energy functionals. Typical examples are the theories of homogenization, of singular perturbations, and of the limit behaviour of elliptic problems with varying obstacles, which have been developed in many papers published in the last fiftheen years (we refer the interested reader in the books by Attouch [21], Bensoussan & Lions & Papanicolaou [44], and Sanchez-Palencia [255] for a systematic presentation and wide bibliographies).

CHAPTER 1

Direct Method and Relaxation. The Abstract Formulation

In this chapter we describe the direct method and the relaxation method in a general topological framework; particular attention is paid to the minimum properties of functionals. In the last section of the chapter some classical examples from the calculus of variations are considered.

1.1. Topological Generalities

Let (X,τ) be a topological space and let $F:X\to\mathbf{R}$ be a mapping.

DEFINITION 1.1.1. *We say that* F *is* τ*-lower semicontinuous (shortly* τ*-l.s.c.) if for every* $t\in\mathbf{R}$ *the set* $\{x\in X : F(x)\le t\}$ *is closed in* X.
We say that F *is sequentially* τ*-lower semicontinuous (shortly seq.* τ*-l.s.c.) if for every sequence* (x_h) τ*-converging to some point* $x\in X$ *it is*

$$F(x) \le \liminf_{h\to\infty} F(x_h) \ .$$

The properties below follow straightforward from Definition 1.1.1.

PROPOSITION 1.1.2. *Let* (X,τ) *be a topological space. Then*

(i) $F: X \to \overline{\mathbf{R}}$ *is* τ*-l.s.c. if and only if the epigraph of* F

$$\mathrm{epi}(F) = \{(x,t) \in X \times \overline{\mathbf{R}} : F(x) \leq t\}$$

is closed in $X \times \overline{\mathbf{R}}$.

(ii) *if* $(F_i)_{i \in I}$ *is a family of* τ*-l.s.c. functionals, then the functional*

$$F(x) = \sup\{F_i(x) : i \in I\}$$

is τ*-l.s.c.;*

(iii) *if* F *and* G *are* τ*-l.s.c. functionals from* X *into* $]-\infty,+\infty]$, *then* $F+G$ *is* τ*-l.s.c..*

DEFINITION 1.1.3. *We say that a subset* E *of* X *is* τ*-compact if for every open covering* $(A_i)_{i \in I}$ *of* E *there exists a finite subset* J *of* I *such that* $(A_j)_{j \in J}$ *is still a covering of* E.

We say that E *is sequentially* τ*-compact if for every sequence* (x_h) *in* E *there exists a subsequence* $(x_{h(k)})$ τ*-converging to an element* $x \in E$.

REMARK 1.1.4. The definition of τ-lower semicontinuity and of τ-compactness can be reformulated in terms of nets; in fact we have (see for instance Kelley [186]):

(i) F is τ-l.s.c. if and only if for every net $(x_i)_{i \in I}$ τ-converging to some point $x \in X$ we have

$$F(x) \leq \liminf_{i \in I} F(x_i) \ ;$$

(ii) $E \subset X$ is τ-compact if and only if for every net $(x_i)_{i \in I}$ in E there exists a subnet $(x_{i(j)})_{j \in J}$ τ-converging to an element $x \in E$.

The following proposition shows that the sequential τ-lower semicontinuity is a topological concept (see for instance Dolcher [140]).

PROPOSITION 1.1.5. *Denote by* τ_{seq} *the topology on* X *whose closed sets are the sequentially* τ*-closed subsets of* X*. Then*

(i) τ_{seq} *is the strongest topology on* X *for which the converging sequences are the* τ*-converging sequences;*

(ii) F *is seq.* τ*-l.s.c. if and only if* F *is* τ_{seq}-l.s.c.*;*

(iii) $\tau_{seq}=\tau$ *for every first countable topology* τ.

For convex functions in a Banach space the following result holds (see Dunford & Schwartz [144], Chapter V).

PROPOSITION 1.1.6. *Let* X *be a Banach space and let* $F:X\to]-\infty,+\infty]$ *be a convex function. Then*

(i) F *is strongly-l.s.c.* $\Leftrightarrow$ F *is weakly-l.s.c.;*

(ii) *if* X' *is separable*

F *is weakly-l.s.c.* $\Leftrightarrow$ F *is seq. weakly-l.s.c.;*

(iii) *if* X=V' *with* V *separable Banach space*

F *is weakly*-l.s.c.* $\Leftrightarrow$ F *is seq. weakly*-l.s.c.* .

Concerning the relations between compactness and sequential compactness it is well-known that if (X,τ) is a metric space, then the two notions actually coincide. In the case of a Banach space endowed with its weak topology, we recall the Eberlein-Smulian theorem (see Dunford & Schwartz [144], page 430).

THEOREM 1.1.7. *Let* X *be a Banach space and let* E *be a subset of* X*. Then the following conditions are equivalent.*

(i) *every sequence in* E *admits a weakly converging subsequence;*

(ii) *the weak closure of* E *is weakly compact.*

1.2. The Direct Method

In this section we introduce the so-called direct method for proving existence results for minima of functionals. Taking into account Proposition 1.1.5, Proposition 1.1.6, and Theorem 1.1.7, we may limit ourselves to consider the direct method only in a topological framework.

As in the previous section let (X,τ) be a topological space and let $F:X\to\overline{\mathbf{R}}$ be a function.

DEFINITION 1.2.1. *We say that* F *is* τ*-coercive if for every* $t\in\mathbf{R}$ *there exists a* τ*-compact (and* τ*-closed) subset* K_t *of* X *such that*

$$\{x\in X : F(x)\le t\} \subset K_t .$$

With the definition above, the direct method consists in the following proposition.

PROPOSITION 1.2.2. *Assume that*

(i) F *is* τ*-l.s.c.;*

(ii) F *is* τ*-coercive.*

Then F *admits a minimum point on* X.

Proof. If F is the constant $+\infty$ the assertion is trivial. Otherwise, it is

$$\inf\{F(x) : x\in X\} = M < +\infty . \tag{1.2.1}$$

By hypotheses (i) and (ii) for every $M'>M$ the set $\{x\in X : F(x)\leq M'\}$ is τ-compact; moreover it is nonempty because of (1.2.1). Then the set

$$\{x\in X \ : \ F(x)\leq M\} = \bigcap_{M'>M} \{x\in X \ : \ F(x)\leq M'\}$$

is nonempty, and this achieves the proof. ∎

REMARK 1.2.3. Proposition 1.2.2 holds, with a similar proof, if in the definition of τ-coerciveness the subsets K_t are only supposed countably compact, in the sense that every countable open covering of K_t has a finite subcovering. Moreover, in the sequential case, the hypotheses

(i) F *is seq.* τ*-l.s.c.,*

(ii) $\forall t\in \mathbf{R}$ $\exists K_t$ *sequentially* τ*-compact with* $\{x\in X : F(x)\leq t\}\subset K_t$,

still imply the existence of a minimum point for F.

The following example is classic in the calculus of variations.

EXAMPLE 1.2.4. Let Ω be a bounded open subset of $\mathbf{R}^n$ with a Lipschitz boundary, and let $X=H^1_o(\Omega)$ be the usual Sobolev space of functions in $L^2(\Omega)$ with first distribution derivatives in $L^2(\Omega)$ and with null trace on the boundary $\partial\Omega$. Consider the classical Dirichlet minimization problem

$$\min\left\{\int_\Omega |Du|^2\,dx + \int_\Omega g(x)\,u\,dx \ : \ u\in H^1_0(\Omega)\right\} \tag{1.2.2}$$

where $g\in L^2(\Omega)$ is a given function.

By Proposition 1.2.2, problem (1.2.2) has a solution. In fact, taking as τ the strong topology of $L^2(\Omega)$, the functional

$$F(u) = \int_\Omega |Du|^2\, dx + \int_\Omega g(x)\, u\, dx$$

turns out to be τ-l.s.c.. To prove the τ-coerciveness, note that by Poincaré inequality, there exists a positive constant c such that for every $u \in X$

$$\int_\Omega u^2\, dx \le c \int_\Omega |Du|^2\, dx\ .$$

Then

$$F(u) \ge \int_\Omega |Du|^2\, dx - \|g\|_{L^2(\Omega)} \|u\|_{L^2(\Omega)} \ge$$

$$\ge \int_\Omega |Du|^2\, dx - \frac{1}{2c}\int_\Omega u^2\, dx - \frac{c}{2}\int_\Omega g^2\, dx \ge$$

$$\ge \frac{1}{2}\int_\Omega |Du|^2\, dx - \frac{c}{2}\int_\Omega g^2\, dx\ ,$$

so that

$$\{u \in X\ :\ F(u) \le t\} \subset \{u \in X\ :\ \int_\Omega |Du|^2\, dx \le 2t + c\int_\Omega g^2\, dx\}$$

which is τ-compact by Rellich Theorem. ∎

REMARK 1.2.5. By an argument similar to the one used in Example 1.2.4 we can prove that a functional F defined on the Sobolev space $W^{1,p}(\Omega)$ (with $p>1$) is coercive with respect to the $L^p(\Omega)$ topology (and also with respect to the weak topology of $W^{1,p}(\Omega)$) provided F satisfies an inequality of the form

$$F(u) \ge \alpha \int_\Omega \left[|Du|^p + |u|^p\right] dx - \beta$$

for suitable $\alpha>0$ and $\beta\geq 0$.

In dual Banach spaces (in particular in reflexive spaces) the coerciveness with respect to the weak* topology is expressed in the following proposition.

PROPOSITION 1.2.6. *Let* $X=V'$ *be the dual of a Banach space* V, *and let* τ *be its weak* topology. Then a function* $F:X\to\overline{\mathbf{R}}$ *is* τ*-coercive if and only if*

$$\lim_{\|x\|\to+\infty} F(x) = +\infty \ . \tag{1.2.3}$$

Proof. Assume (1.2.3) holds; then for every $t\in\mathbf{R}$ there exists $M_t>0$ such that

$$\|x\|>M_t \ \Rightarrow \ F(x)>t \, ,$$

so that the set $K_t=\{x\in X : \|x\|\leq M_t\}$ is τ-compact and

$$\{x\in X : F(x)\leq t\} \subset K_t \, .$$

Assume (1.2.3) does not hold; then there exist $t\in\mathbf{R}$ and a sequence (x_h) in X with $\|x_h\|>h$ and $F(x_h)\leq t$, and this contradicts the definition of τ-coerciveness. ■

REMARK 1.2.7. It is immediately seen that condition (1.2.3) is equivalent to the following one:

there exists a function $\Phi:\mathbf{R}\to\mathbf{R}$ *such that*

$$\lim_{s\to+\infty} \Phi(s) =+\infty \qquad \textit{and} \qquad F(x) \geq \Phi(\|x\|) \, .$$

In order to characterize the weak-coercive functionals $F:L^1(\Omega)\to\overline{\mathbf{R}}$ we recall the Dunford-Pettis compactness theorem (see for instance Dellacherie & Meyer [138], pages 24 and 27).

THEOREM 1.2.8. *Let* $(\Omega,\mathfrak{I},\mu)$ *be a measure space with* μ *positive and finite, and let* $\mathbb{K}$ *be a subset of* $L^1(\Omega,\mathfrak{I},\mu)$. *Then, the following conditions are equivalent.*

(a) $\mathbb{K}$ *is relatively compact in the weak topology of* $L^1(\Omega,\mathfrak{I},\mu)$;

(b) $\mathbb{K}$ *is uniformly integrable, that is* $\mathbb{K}$ *is bounded in* $L^1(\Omega,\mathfrak{I},\mu)$ *and for every* $\varepsilon>0$ *there exists* $\delta>0$ *such that*

$$A\in\mathfrak{I}\ ,\quad \mu(A)<\delta \quad\Rightarrow\quad \sup\Big\{\int_A |u|\,d\mu \ :\ u\in\mathbb{K}\Big\} < \varepsilon\ ;$$

(c) *there exists a function* $\theta:[0,+\infty[\to[0,+\infty[$ *(*θ *can be taken convex and increasing) such that*

$$\lim_{s\to+\infty}\frac{\theta(s)}{s} = +\infty \qquad \textit{and} \qquad \sup\Big\{\int_\Omega \theta(|u|)\,d\mu \ :\ u\in\mathbb{K}\Big\} < +\infty\ .$$

PROPOSITION 1.2.9. *Let* $(\Omega,\mathfrak{I},\mu)$ *be as in Theorem 1.2.8 and let* $F:L^1(\Omega,\mathfrak{I},\mu)\to\overline{\mathbf{R}}$ *be a functional. Then* F *is weakly-coercive if and only if*

$$F(u) \geq \Phi\Big(\int_\Omega \theta(|u|)\,d\mu\Big) \qquad \textit{for every } u\in L^1(\Omega,\mathfrak{I},\mu) \tag{1.2.4}$$

for suitable functions $\Phi:\mathbf{R}\to\mathbf{R}$ *and* $\theta:\mathbf{R}\to\mathbf{R}$ *such that*

$$\lim_{s\to+\infty}\Phi(s) = +\infty \qquad \textit{and} \qquad \lim_{s\to+\infty}\frac{\theta(s)}{s} = +\infty\ .$$

Proof. If (1.2.4) holds, then F is weakly-coercive by Dunford-Pettis theorem. On the contrary, assume F is weakly-coercive; by Dunford-Pettis theorem again, we have that for every $h\in\mathbf{N}$ there exists a function $\theta_h:\mathbf{R}\to\mathbf{R}$ such that

$$\lim_{s\to+\infty}\frac{\theta_h(s)}{s} = +\infty \qquad \text{and} \qquad F(u)\leq h \ \Rightarrow\ \int_\Omega \theta_h(|u|)\,d\mu \leq 1\ .$$

Moreover, $\theta_h(s)$ can be taken decreasing with respect to h and increasing with respect to s. It is easy to see that (1.2.4) is proved if we construct a sequence (M_h) in $]0,+\infty[$ and a function $\theta:\mathbf{R}\to\mathbf{R}$ such that

$$(1.2.5) \qquad \lim_{s\to+\infty}\frac{\theta(s)}{s}=+\infty \qquad \text{and} \qquad F(u)\le h \;\Rightarrow\; \int_\Omega \theta(|u|)\,d\mu \le M_h\,.$$

For every sequence $R_h\to+\infty$ (with $R_0=0$) we define

$$\theta(s) = \theta_h(s) \qquad \text{if } s\in[R_h,R_{h+1}[$$

$$M_h = 1+\mu(\Omega)\sum_{i<h}\theta_i(R_{i+1})\ .$$

It is possible to choose the sequence (R_h) such that

$$\lim_{s\to+\infty}\frac{\theta(s)}{s} = +\infty\ .$$

Moreover, if $F(u)\le h$ we have

$$\int_\Omega \theta(|u|)\,d\mu = \int_{|u|\ge R_h}\theta(|u|)\,d\mu + \sum_{i<h}\int_{R_i\le|u|<R_{i+1}}\theta(|u|)\,d\mu \le$$

$$\le \int_\Omega \theta_h(|u|)\,d\mu + \sum_{i<h}\int_{R_i\le|u|<R_{i+1}}\theta_i(|u|)\,d\mu \le$$

$$\le 1+\mu(\Omega)\sum_{i<h}\theta_i(R_{i+1}) = M_h\ ,$$

so that (1.2.5) is proved. ■

1.3. Relaxation

When condition (i) of Proposition 1.2.2 is not fulfilled, in general the existence of minimum points for F may fail. However, it is interesting to study the behaviour of the minimizing sequences of F (which, under hypothesis (ii), are τ-compact). To this aim, we introduce the τ-*relaxed* functional of F given by

$$\Gamma(\tau^-)F(x) = \sup \{G(x) : G \text{ is } \tau\text{-l.s.c.}, G \leq F\} .$$

Note that by Proposition 1.1.2(ii) the supremum above is actually attained, so that $\Gamma(\tau^-)F$ turns out to be the greatest τ-l.s.c. functional less than or equal to F.

The properties of the functional $\Gamma(\tau^-)F$ are listed in the following proposition.

PROPOSITION 1.3.1. *The following properties hold:*

(i) $\Gamma(\tau^-)F$ *is a* τ-*l.s.c. functional;*

(ii) *for every* $x \in X$ *it is*

$$\Gamma(\tau^-)F(x) = \liminf_{y \to x} F(y) =$$

$$= \min \left\{ \liminf_{i \in I} F(x_i) : (x_i)_{i \in I} \textit{ is a net } \tau\textit{-converging to } x \right\} ;$$

(iii) *for every* $x \in X$ *we have*

$$\inf \{F(y) : y \in X\} \leq \Gamma(\tau^-)F(x) \leq F(x) ;$$

in particular it is

$$\inf \{F(x) : x \in X\} = \inf \{\Gamma(\tau^-)F(x) : x \in X\} ;$$

(iv) *for every* F *which is* τ-*coercive,* $\Gamma(\tau^-)F$ *is* τ-*coercive;*

(v) *for every* τ-*continuous function* $G : X \to \mathbf{R}$ *it is*

$$\Gamma(\tau^-)[G+F] = G + \Gamma(\tau^-)F ;$$

(vi) *if* (x_h) *is a minimizing sequence for* F, *that is*

$$\lim_{h\to\infty} F(x_h) = \inf\{F(y) : y\in X\} ,$$

which τ*-converges to some point* $x\in X$, *then* x *is a minimum point for* $\Gamma(\tau^-)F$.

Proof. Assertion (i) follows from Proposition 1.1.2(ii).

Assertion (ii) follows from the fact that the functional

$$x \to \liminf_{y\to x} F(y)$$

is τ-l.s.c., less than or equal to F, and such that for every τ-l.s.c. functional $G\leq F$ we have

$$G(x) = \liminf_{y\to x} G(y) \leq \liminf_{y\to x} F(y) .$$

To prove assertion (iii) it suffices to remark that the constant functional

$$G(x) = \inf\{F(y) : y\in X\}$$

is τ-l.s.c. and less than or equal to F.

We prove assertion (iv). If F is τ-coercive, for every $t\in\mathbf{R}$ we have

$$\{x\in X : F(x)\leq t\} \subset K_t$$

for suitable τ-compact and τ-closed subsets K_t of X. We claim that

$$\{x\in X : \Gamma(\tau^-)F(x)\leq t\} \subset K_{t+1}$$

for every $t\in\mathbf{R}$. In fact, let $\Gamma(\tau^-)F(x)\leq t$; by assertion (ii) for every τ-neighbourhood A of x there exists a point $y\in A$ such that $F(y)<t+1$. This means that every τ-neighbourhood of x meets the set K_{t+1}, hence $x\in K_{t+1}$ because K_{t+1} is closed.

The proof of assertion (v) is trivial.

We prove assertion (vi). Since $\Gamma(\tau^-)F$ is τ-l.s.c., by using assertion (iii) it is

$$\Gamma(\tau^-)F(x) \leq \liminf_{h\to\infty} \Gamma(\tau^-)F(x_h) \leq$$

$$\le \liminf_{h\to\infty} F(x_h) = \inf\{F(y) : y\in X\} =$$

$$= \inf\{\Gamma(\tau^-)F(y) : y\in X\} ,$$

and so x is a minimum point for $\Gamma(\tau^-)F$. ■

When the topology τ is substituted by its sequential version τ_{seq}, the τ_{seq}-relaxed functional $\Gamma((\tau_{seq})^-)F$ is usually indicated by $\Gamma_{seq}(\tau^-)F$. This functional is very useful in the applications, when we deal with weak topologies of L^p spaces or Sobolev spaces. The following proposition gives a characterization of the functional $\Gamma_{seq}(\tau^-)F$.

PROPOSITION 1.3.2. *Set for every* $x\in X$

$$(1.3.1) \qquad S_\tau F(x) = \inf\Big\{\liminf_{h\to\infty} F(x_h) \ :\ (x_h) \textit{ is a sequence } \tau\textit{-converging to } x\Big\}.$$

Let Λ *be the set of all countable ordinals; for every* $\lambda\in\Lambda$ *we define a functional* F_λ *by transfinite induction:*

$$F_0 = F$$

$$F_{\lambda+1} = S_\tau F_\lambda$$

$$F_\lambda = \inf\{F_\mu : \mu<\lambda\} \qquad \textit{if } \lambda \textit{ is a limit ordinal} .$$

Then we have

$$\Gamma_{seq}(\tau^-)F = \inf\{F_\lambda : \lambda\in\Lambda\} .$$

Proof. It is clear that $F_\lambda\le F_\mu$ if $\lambda\ge\mu$; hence, setting

$$H(x) = \inf\{F_\lambda(x) : \lambda\in\Lambda\} ,$$

we have $H\le F$. Suppose now that G is a seq. τ-l.s.c. functional with $G\le F$; it is easy to prove, by transfinite induction, that $G\le F_\lambda$ for every $\lambda\in\Lambda$, hence $G\le H$.

To conclude the proof, it remains to prove that H is seq. τ-l.s.c.. Let (x_h) be a sequence in X which τ-converges to some $x\in X$; for every $h\in\mathbf{N}$ there exists $\lambda_h\in\Lambda$ such that

$$F_{\lambda_h}(x_h) \le H(x_h) + \frac{1}{h} . \tag{1.3.2}$$

Let $\lambda_\infty=\sup\{\lambda_h : h\in\mathbf{N}\}$; then $\lambda_\infty\in\Lambda$ and by (1.3.2) we get

$$\begin{aligned} H(x) &\le F_{\lambda_\infty+1}(x) = S_\tau F_{\lambda_\infty}(x) \le \\ &\le \liminf_{h\to\infty} F_{\lambda_\infty}(x_h) \le \liminf_{h\to\infty} F_{\lambda_h}(x_h) \le \\ &\le \liminf_{h\to\infty} \left[H(x_h) + \frac{1}{h}\right] = \liminf_{h\to\infty} H(x_h) . \blacksquare \end{aligned}$$

In the following propositions we consider other cases, occurring very often in the applications, in which $\Gamma_{seq}(\tau^-)F=\Gamma(\tau^-)F$.

PROPOSITION 1.3.3. *Assume that the point $x\in X$ has a countable fundamental system of τ-neighbourhoods. Then*

$$\begin{aligned} \Gamma(\tau^-)F(x) &= \Gamma_{seq}(\tau^-)F(x) = \\ &= \min\left\{\liminf_{h\to\infty} F(x_h) : (x_h) \textit{ is a sequence } \tau\textit{-converging to } x\right\} . \end{aligned}$$

Proof. Since $\Gamma(\tau^-)F(x)\le S_\tau F(x)$, in order to get the equality $\Gamma(\tau^-)F(x)\ge S_\tau F(x)$ we have only to prove that there exists a sequence (x_h) τ-converging to x such that

$$\liminf_{h\to\infty} F(x_h) \le \Gamma(\tau^-)F(x) . \tag{1.3.3}$$

If the right-hand side is $+\infty$ this is trivial; otherwise, let $U_1,\dots,U_h,\dots$ be a fundamental system of τ-neighbourhoods of x in X. For every $h\in\mathbf{N}$ there exists $x_h\in U_h$

such that

$$F(x_h) \leq \frac{1}{h} + \Gamma(\tau^-)F(x) \ ;$$

hence $x_h \to x$ in X and (1.3.3) follows. Finally, the fact that the infimum in the definition of S_τ is actually a minimum, can be proved by a standard diagonalization argument. ■

PROPOSITION 1.3.4. *Let* X *be a Banach space, let* $F:X\to]-\infty,+\infty]$ *be a convex function, and let one of the following conditions be satisfied:*

(i) X' *is separable and* τ *is the weak topology on* X;

(ii) X=V' *with* V *separable Banach space, and* τ *is the weak* topology on* X.

Then

$$\Gamma(\tau^-)F = \Gamma_{seq}(\tau^-)F \ .$$

Proof. We give the proof only in the case (i), the case (ii) being similar. Since τ_{seq} is a topology stronger than τ, we have

$$\Gamma(\tau^-)F \leq \Gamma_{seq}(\tau^-)F \ .$$

To prove the opposite inequality it is enough to show that $\Gamma_{seq}(\tau^-)F$ is τ-l.s.c.. Since F is convex, it is not difficult to see that $\Gamma_{seq}(\tau^-)F$ is convex too, and so, by Proposition 1.1.6(ii), $\Gamma_{seq}(\tau^-)F$ is τ-l.s.c. because it is τ_{seq}-l.s.c.. ■

PROPOSITION 1.3.5. *Let* (X,τ) *be a topological space and let* $F:X\to\overline{\mathbf{R}}$ *be a function. Assume that*

(i) *every* τ*-compact subset of* X *is metrizable;*

(ii) F *is* τ*-coercive in the sense of Definition 1.2.1.*

Then, if $S_\tau F$ *is the functional defined in (1.3.1), we have*

$$\Gamma(\tau^-)F = \Gamma_{seq}(\tau^-)F = S_\tau F .$$

Proof. The inequalities

$$\Gamma(\tau^-)F \le \Gamma_{seq}(\tau^-)F \le S_\tau F$$

are immediate; it remains to prove the inequality

$$S_\tau F \le \Gamma(\tau^-)F .$$

To do this, it is enough to show that $S_\tau F$ is a τ-l.s.c. functional, that is for every $t\in \mathbf{R}$ the set

$$C_t = \{x\in X : S_\tau F(x)\le t\}$$

is τ-closed. Since F is τ-coercive, the functional $\Gamma(\tau^-)F$ is τ-coercive too (see Proposition 1.3.1(iv)); then, since $S_\tau F\ge\Gamma(\tau^-)F$, also $S_\tau F$ is τ-coercive. Therefore, the set C_t is contained in a τ-compact subset of X, and by hypothesis (i) it is enough to show that C_t is sequentially τ-closed. Let (x_h) be a sequence in C_t τ-converging to x; by the definition of $S_\tau F$, for every $h\in \mathbf{N}$ there exists a sequence $(x_{h,k})$ τ-converging to x_h as $k\to+\infty$ such that

$$F(x_{h,k}) \le t + \frac{1}{h} \qquad \text{for every } k\in \mathbf{N} .$$

By hypothesis (ii), for every $s\in \mathbf{R}$ there exists a τ-compact subset K_s of X such that

$$\{y\in X : F(y)\le s\} \subset K_s ,$$

hence we have

$$x_{h,k}\in K_{t+1} \qquad \text{for every } h,k\in \mathbf{N} .$$

Since K_{t+1} is metrizable, by a standard diagonalization argument we may find a sequence (y_h) τ-converging to x such that

$$F(y_h) \leq t + \frac{1}{h} \qquad \text{for every } h \in \mathbf{N} .$$

Thus

$$S_\tau F(x) \leq \liminf_{h \to \infty} F(y_h) \leq t ,$$

so that $x \in C_t$. ■

REMARK 1.3.6. (See Dunford & Schwartz [144], page 434). Hypothesis (i) of Proposition 1.3.5 is fulfilled either if X is a separable Banach space and τ is the weak topology on X or $X=V'$ with V separable Banach space and τ is the weak* topology on X.

Let us now define the Moreau-Yosida transform F_λ of a given functional F. Let X be a metric space with distance d, and let $F:X \to [0,+\infty]$ be a functional. For every $\lambda > 0$ we define a functional F_λ by setting

$$F_\lambda(x) = \inf \{F(y) + \lambda d(x,y) : y \in X\} .$$

PROPOSITION 1.3.7. *The functionals* F_λ *satisfy the following properties:*

(i) *for every* $\lambda > 0$ *the functional* F_λ *is* λ*-Lipschitz continuous, that is*

$$F_\lambda(x) \leq F_\lambda(y) + \lambda d(x,y) \qquad \textit{for every } x,y \in X;$$

(ii) *for every* $x \in X$ *we have*

$$\lim_{\lambda \to +\infty} F_\lambda(x) = \Gamma(X^-)F(x) .$$

Proof. We prove (i). Let $x,y,z \in X$ and let $\lambda > 0$; the triangle inequality yields

$$F(z) + \lambda d(x,z) \leq F(z) + \lambda d(y,z) + \lambda d(x,y) ,$$

and passing to the infimum for $z\in X$, the proof of (i) is achieved.

We prove (ii). Since by (i) F_λ is Lipschitz continuous, and $F_\lambda \le F$, we have $F_\lambda \le \Gamma(X^-)F$, so that

$$\sup\{F_\lambda(x) : \lambda>0\} \le \Gamma(X^-)F(x) \qquad \text{for every } x\in X .$$

Fix now $x\in X$ and assume that

$$\sup\{F_\lambda(x) : \lambda>0\} < +\infty .$$

Let $x_\lambda \in X$ be such that

$$F_\lambda(x) + \frac{1}{\lambda} \ge F(x_\lambda) + \lambda d(x,x_\lambda) \ ;$$

then $x_\lambda \to x$ as $\lambda\to+\infty$, and so

$$\Gamma(X^-)F(x) \le \liminf_{\lambda\to+\infty} F(x_\lambda) \le$$

$$\le \liminf_{\lambda\to+\infty} F(x_\lambda) + \lambda d(x,x_\lambda) \le \liminf_{\lambda\to+\infty} F_\lambda(x) \ ,$$

which concludes the proof of (ii). ■

1.4. Some Examples

In the calculus of variations the relaxation method can be used to define in a correct way the minimum problems on the appropriated spaces. In the following examples we consider the classical Dirichlet integral and the area-integral; by Ω we denote a bounded open subset of $\mathbf{R}^n$ with a Lipschitz boundary.

EXAMPLE 1.4.1. Let $X=L^2(\Omega)$, let τ be the strong $L^2(\Omega)$ topology, and let

$F:X\to[0,+\infty]$ be the functional defined by

$$F(u) = \begin{cases} \int_\Omega |Du|^2\,dx & \text{if } u\in C^1(\Omega) \\ +\infty & \text{otherwise .} \end{cases}$$

The functional F can be extended, via the relaxation method, to a new functional $\Gamma(\tau^-)F$ whose domain is the Sobolev space $H^1(\Omega)$. Indeed, by using the well-known result that $C^1(\Omega)$ is dense in $H^1(\Omega)$ with respect to the $H^1(\Omega)$-norm

$$\|u\|_{H^1(\Omega)} = \|u\|_{L^2(\Omega)} + \|Du\|_{L^2(\Omega)}$$

(see for instance Meyers & Serrin [224] and Adams [6], Theorem 3.16), it is easy to obtain that

$$H^1(\Omega) = \{u\in X : \Gamma(\tau^-)F(u)<+\infty\}$$

$$\Gamma(\tau^-)F(u) = \begin{cases} \int_\Omega |Du|^2\,dx & \text{if } u\in H^1(\Omega) \\ +\infty & \text{otherwise .} \end{cases}$$

Consider now, with the same choice of X and τ, the functional with "null trace constraint"

$$F_0(u) = \begin{cases} \int_\Omega |Du|^2\,dx & \text{if } u\in C^1_0(\Omega) \\ +\infty & \text{otherwise .} \end{cases}$$

In this case we obtain

$$H^1_0(\Omega) = \{u\in X : \Gamma(\tau^-)F_0(u)<+\infty\}$$

$$\Gamma(\tau^-)F_0(u) = \begin{cases} \int_\Omega |Du|^2\,dx & \text{if } u\in H^1_0(\Omega) \\ +\infty & \text{otherwise .} \end{cases}$$

Note that by Proposition 1.3.1(v) we have for every $g\in L^2(\Omega)$

$$\Gamma(\tau^-)\Big[F(u) + \int_\Omega g(x)\,u\,dx\Big] = \Big[\Gamma(\tau^-)F(u)\Big] + \int_\Omega g(x)\,u\,dx$$

$$\Gamma(\tau^-)\Big[F_0(u) + \int_\Omega g(x)\, u\, dx\Big] = \Big[\Gamma(\tau^-)F_0(u)\Big] + \int_\Omega g(x)\, u\, dx\ .$$

Moreover, by Proposition 1.3.1(vi), the Poincaré inequality, and the Rellich compactness theorem, every minimizing sequence of the problem

$$\inf\Big\{\int_\Omega |Du|^2\, dx + \int_\Omega g(x)\, u\, dx\ :\ u\in C^1_0(\Omega)\Big\}$$

converges in $L^2(\Omega)$ to the solution of

$$\min\Big\{\int_\Omega |Du|^2\, dx + \int_\Omega g(x)\, u\, dx\ :\ u\in H^1_0(\Omega)\Big\}\ .$$

EXAMPLE 1.4.2. Let $X=L^1(\Omega)$, let τ be the strong $L^1(\Omega)$ topology, and let $F:X\to[0,+\infty]$ be the functional defined by

$$\text{(1.4.1)}\qquad F(u) = \begin{cases} \int_\Omega \sqrt{1+|Du|^2}\, dx & \text{if } u\in C^1(\Omega) \\ +\infty & \text{otherwise}\ . \end{cases}$$

In this case, the relaxation method leads us to introduce the space of functions with bounded variation on Ω

$$BV(\Omega) = \{u\in L^1(\Omega)\ :\ Du \text{ is a measure with finite total variation on } \Omega\}\ .$$

It is well known that the space $C^1(\Omega)$ is not dense in $BV(\Omega)$ with respect to the $BV(\Omega)$-norm

$$\|u\|_{BV(\Omega)} = \|u\|_{L^1(\Omega)} + \int_\Omega |Du|$$

(in fact the closure of $C^1(\Omega)$ in the norm above is the space $W^{1,1}(\Omega)$); nevertheless, it is possible to prove (see Anzellotti & Giaquinta [18]) that $C^1(\Omega)$ is dense in $BV(\Omega)$ with respect to the distance

$$d(u,v) = \|u-v\|_{L^1(\Omega)} + \left| \int_\Omega |Du| - \int_\Omega |Dv| \right| ,$$

and this allows to get

$$BV(\Omega) = \{ u \in X : \Gamma(\tau^-)F(u) < +\infty \}$$

$$\Gamma(\tau^-)F(u) = \begin{cases} \int_\Omega \sqrt{1+|R(Du)|^2}\, dx + |S(Du)|(\Omega) & \text{if } u \in BV(\Omega) \\ +\infty & \text{otherwise ,} \end{cases}$$

where $Du = R(Du)dx + S(Du)$ is the usual Lebesgue decomposition of the measure Du into the absolute continuous part $R(Du)dx$ and the singular part $S(Du)$ with respect to the Lebesgue measure, and $|S(Du)|(\Omega)$ denotes the total variation of the measure $S(Du)$ on Ω.

If we consider, as in Example 1.4.1, the functional with "null trace constraint"

$$(1.4.2) \qquad F_0(u) = \begin{cases} \int_\Omega \sqrt{1+|Du|^2}\, dx & \text{if } u \in C^1_0(\Omega) \\ +\infty & \text{otherwise ,} \end{cases}$$

we may remark a very deep difference between this case and the analogous case of Example 1.4.1. In fact, the domain of the relaxed functional $\Gamma(\tau^-)F_0$ is still $BV(\Omega)$, and we get (see for instance Giusti [170], Miranda [229])

$$\Gamma(\tau^-)F_0(u) = \begin{cases} \int_\Omega \sqrt{1+|R(Du)|^2}\, dx + |S(Du)|(\Omega) + \int_{\partial\Omega} |\gamma(u)|\, dH^{n-1} & \text{if } u \in BV(\Omega) \\ +\infty & \text{otherwise,} \end{cases}$$

where H^{n-1} denotes the n–1 dimensional Hausdorff measure and $\gamma(u) \in L^1(\partial\Omega)$ is the boundary trace of u (see Anzellotti & Giaquinta [18]). In other words, for the area-integrals (1.4.1) and (1.4.2) the domain of the relaxed functionals is in both cases $BV(\Omega)$, and the "null trace constraint" in (1.4.2) relaxes to the boundary penalization

$$\int_{\partial\Omega} |\gamma(u)|\, dH^{n-1} .$$

For further extensions of this result to more general integrands and boundary terms, we refer the interested reader to some recent papers by Carriero & Leaci & Pascali (see [84], [85], [86]).

In the following example a case of an integral functional depending on vector-valued functions is considered.

EXAMPLE 1.4.3. Let $X=L^p(\Omega;\mathbf{R}^n)$ $(n\geq 2,\ 1\leq p\leq+\infty)$, let τ be the strong $L^p(\Omega;\mathbf{R}^n)$ topology, and let $F:X\to[0,+\infty]$ be the functional defined by

$$(1.4.3) \qquad F(u) = \begin{cases} \displaystyle\int_\Omega |\det Du|\, dx & \text{if } u\in C^1(\Omega;\mathbf{R}^n) \\ +\infty & \text{otherwise} . \end{cases}$$

If $p<+\infty$ we have (see Acerbi & Buttazzo [2])

$$(1.4.4) \qquad \Gamma(\tau^-)F(u) = 0 \qquad \text{for every } u\in L^p(\Omega;\mathbf{R}^n) .$$

Proof. By a density argument it is enough to prove (1.4.4) for every smooth function. Let u be such a function; we will find a sequence (u_h) in $C^1(\Omega;\mathbf{R}^n)$ such that

$$(1.4.5) \qquad u_h \to u \ \text{ in } L^p(\Omega;\mathbf{R}^n) \qquad \text{and} \qquad \det Du_h = 0 \ \text{ in } \Omega .$$

Let Y be the unitary cube $Y=]0,1[^n$ and, for all $h\in\mathbf{N}$ set

$$G_h = \{h^{-1}(x+Y) : x\in\mathbf{Z}^n\}$$

$$T_h = \{Q\in G_h : Q\subset\Omega\}$$

$$\Omega_h = \bigcup_{Q\in T_h} Q .$$

If $u^{(i)}$ denotes the i-th component of u $(i=1,...,n)$, for every $Q \in T_h$ set

$$u_Q^{(i)} = (\text{meas } Q)^{-1} \int_Q u^{(i)} \, dx \ .$$

Finally, let (u_h) be a sequence in $C^1(\Omega;\mathbf{R}^n)$ such that on every $Q \in T_h$ it is

$$u_h^{(i)}(x) = 0 \qquad \text{if } \operatorname{dist}(x,\partial Q) \le ih^{-2}$$

$$u_h^{(i)}(x) = u_Q^{(i)} \qquad \text{if } \operatorname{dist}(x,\partial Q) \ge (i+1)h^{-2}$$

$$|u_h^{(i)}(x)| \le |u_Q^{(i)}| \qquad \text{for every } x \in Q \ ,$$

and

$$u_h(x) = 0 \qquad \text{if } x \notin \Omega_h \ .$$

It is easy to see that $\det Du_h = 0$ in Ω; moreover

$$\lim_{h\to\infty} \int_\Omega |u-u_h|^p \, dx = \lim_{h\to\infty} \sum_{Q\in T_h} \int_Q |u-u_h|^p \, dx =$$

$$= \lim_{h\to\infty} \sum_{Q\in T_h} \int_Q |u-u_Q|^p \, dx = 0 \ ,$$

hence (1.4.5) is proved. ■

REMARK 1.4.4. When τ is the strong $L^\infty(\Omega;\mathbf{R}^n)$ topology and F is given by (1.4.3), it is possible to prove that (we refer to Acerbi & Buttazzo [2] and Acerbi & Buttazzo & Fusco [3], [4] for further details)

$$\Gamma(\tau^-)F(u) = \int_\Omega |\det Du| \, dx \qquad \text{for every } u \in W^{1,n}(\Omega;\mathbf{R}^n) \cap C(\Omega;\mathbf{R}^n) \ .$$

However, the problem of characterizing the set

$$\left\{ u \in C(\Omega;\mathbf{R}^n) \ : \ \Gamma(\tau^-)F(u) < +\infty \right\}$$

and of representing on that set the functional $\Gamma(\tau^-)F$ is still open.

CHAPTER 2

Functionals Defined on L^p Spaces

In this chapter we study the relaxation problem for functionals of the form

$$F(u,v) = \int_\Omega f(x,u(x),v(x))\, d\mu(x) \qquad u\in L^p(\Omega;\mathbf{R}^m),\ v\in L^q(\Omega;\mathbf{R}^n)$$

with respect to the strong topology on the space $L^p(\Omega;\mathbf{R}^m)$ and the weak topology on the space $L^q(\Omega;\mathbf{R}^n)$. To do this, we shall use an integral representation result (Theorem 2.4.6) which consists in showing that every mapping $G:L^p(\Omega;\mathbf{R}^m)\to\ ]-\infty,+\infty]$ satisfying suitable conditions is actually an integral functional of the form

$$G(u) = \int_\Omega g(x,u(x))\, d\mu(x) \ .$$

In all this chapter $(\Omega,\mathfrak{I},\mu)$ will denote a measure space, with μ non-negative and finite. We shall assume that $\mathfrak{I}$ is μ-complete (i.e. $B\in\mathfrak{I}$ whenever $B\subset N$, for some $N\in\mathfrak{I}$ with $\mu(N)=0$), and that μ is non-atomic (i.e. for every $B\in\mathfrak{I}$ with $\mu(B)>0$ there exists $E\subset B$ such that $0<\mu(E)<\mu(B)$).

For every $p\in[1,+\infty]$ and for every integer number $m\geq 1$ we shall denote by $wL^p(\Omega;\mathbf{R}^m)$ the weak topology of $L^p(\Omega;\mathbf{R}^m)$, by $w^*L^\infty(\Omega;\mathbf{R}^m)$ the weak* topology of $L^\infty(\Omega;\mathbf{R}^m)$, and by $\mathbb{B}_m$ the Borel σ-field of $\mathbf{R}^m$.

2.1. Integrands

This section is devoted to the study of properties of $\mathfrak{I}\otimes\mathbb{B}_m$-measurable functions $f:\Omega\times\mathbf{R}^m\to\overline{\mathbf{R}}$.

DEFINITION 2.1.1. *A function* $f:\Omega\times\mathbf{R}^m\to]-\infty,+\infty]$ *is said to be*

(a) *an* **integrand** *if* f *is* $\mathfrak{I}\otimes\mathbb{B}_m$*-measurable;*

(b) *a* **normal integrand** *if* f *is an integrand and* $f(x,\cdot)$ *is l.s.c. on* $\mathbf{R}^m$ *for* μ*-a.e.* $x\in\Omega$;

(c) *a* **convex integrand** *if* f *is an integrand and* $f(x,\cdot)$ *is convex and l.s.c. on* $\mathbf{R}^m$ *for* μ*-a.e.* $x\in\Omega$;

(d) *a* **Carathéodory integrand** *if* f *is an integrand and* $f(x,\cdot)$ *is finite and continuous on* $\mathbf{R}^m$ *for* μ*-a.e.* $x\in\Omega$.

A function $f:\Omega\times\mathbf{R}^m\times\mathbf{R}^n\to]-\infty,+\infty]$ *is said to be*

(e) *a* **normal-convex integrand** *if* f *is* $\mathfrak{I}\otimes\mathbb{B}_m\otimes\mathbb{B}_n$*-measurable and there exists a* μ*-negligible set* $N\subset\Omega$ *such that*

$$f(x,\cdot,\cdot) \text{ is l.s.c. on } \mathbf{R}^m\times\mathbf{R}^n \text{ for every } x\in\Omega-N$$

$$f(x,s,\cdot) \text{ is convex on } \mathbf{R}^n \text{ for every } x\in\Omega-N,\ s\in\mathbf{R}^m.$$

DEFINITION 2.1.2. *Let* f *and* g *be two functions from* $\Omega\times\mathbf{R}^m$ *into* $\overline{\mathbf{R}}$. *We say that*

(i) f *is* μ-**dominated** *by* g *(and we write* $f \lesssim g$ *) if there exists a* μ*-negligible set* $N\subset\Omega$ *such that*

$$f(x,s) \leq g(x,s) \qquad \text{for every } x\in\Omega-N,\ s\in\mathbf{R}^m;$$

(ii) f *and* g *are* μ-**equivalent** *(and we write* $f\approx g$ *) if* $f\lesssim g$ *and* $g\lesssim f$.

PROPOSITION 2.1.3. *Let* f *and* g *be two integrands. Suppose there exists an integrand* $\psi:\Omega\times[0,+\infty[\to[0,+\infty[$ *such that*

(i) $\psi(x,t)$ *is increasing in* t *and* μ*-integrable in* x;

(ii) *the integrand* $-\psi(x,|s|)$ *is* μ*-dominated either by* f *or by* g;

(iii) $\int_B f(x,u(x))\,d\mu(x) \le \int_B g(x,u(x))\,d\mu(x)$ *for every* $u\in L^\infty(\Omega;\mathbf{R}^m)$ *and* $B\in\mathfrak{I}$.

Then we have $f \lesssim g$.

The proof of Proposition 2.1.3 relies on the following measurable selection theorem due to R.I.Aumann (for the proof we refer to Castaing & Valadier [88], Theorems III.22 and III.23).

THEOREM 2.1.4. *Let* $S\in\mathfrak{I}\otimes\mathbb{B}_m$, *for every* $x\in\Omega$ *let* $S(x)=\{s\in\mathbf{R}^m : (x,s)\in S\}$, *and let* $\Omega_0=\{x\in\Omega : S(x)\neq\emptyset\}$. *Then* $\Omega_0\in\mathfrak{I}$ *and there exists an* $\mathfrak{I}$*-measurable function* $\sigma:\Omega_0\to\mathbf{R}^m$ *such that*

$$\sigma(x)\in S(x) \qquad \textit{for every } x\in\Omega_0 .$$

Proof of Proposition 2.1.3. For every $k\in\mathbf{N}$ let $f_k=f\wedge k$ and $g_k=g\wedge k$; it is enough to prove that for every $k\in\mathbf{N}$ we have

(2.1.1) $\qquad f_k(x,s) \le g_k(x,s)$ for μ-a.e. $x\in\Omega$ and for all $s\in\mathbf{R}^m$ with $|s|\le k$.

Define for $\varepsilon>0$

$$S_{\varepsilon,k} = \{(x,s)\in\Omega\times\mathbf{R}^m \ : \ |s|\le k\,,\ f_k(x,s) > g_k(x,s) + \varepsilon\}$$

$$S_{\varepsilon,k}(x) = \{s\in\mathbf{R}^m \ : \ (x,s)\in S_{\varepsilon,k}\}$$

$$\Omega_{\varepsilon,k} = \{x\in\Omega \ : \ S_{\varepsilon,k}(x)\neq\emptyset\} .$$

The set $S_{\varepsilon,k}$ belongs to $\mathfrak{I}\otimes\mathbb{B}_m$; therefore by Theorem 2.1.4 the set $\Omega_{\varepsilon,k}$ belongs to $\mathfrak{I}$ and there exists an $\mathfrak{I}$-measurable function $\sigma_{\varepsilon,k}:\Omega_{\varepsilon,k}\to\mathbf{R}^m$ such that

$\sigma_{\varepsilon,k}(x)\in S_{\varepsilon,k}(x)$ for every $x\in\Omega_{\varepsilon,k}$. Define $\sigma_{\varepsilon,k}(x)=0$ for $x\in\Omega-\Omega_{\varepsilon,k}$; since $|\sigma_{\varepsilon,k}|\le k$, we have $\sigma_{\varepsilon,k}\in L^\infty(\Omega;\mathbf{R}^m)$, so that by hypothesis (iii)

(2.1.2) $$\int_{\Omega_{\varepsilon,k}} f(x,\sigma_{\varepsilon,k}(x))\, d\mu(x) \le \int_{\Omega_{\varepsilon,k}} g(x,\sigma_{\varepsilon,k}(x)\, d\mu(x)\ .$$

By the definition of $S_{\varepsilon,k}$ we have

(2.1.3) $$f_k(x,\sigma_{\varepsilon,k}(x)) > g_k(x,\sigma_{\varepsilon,k}(x)) + \varepsilon \qquad \text{for every } x\in\Omega_{\varepsilon,k}\ ,$$

hence

(2.1.4) $$g_k(x,\sigma_{\varepsilon,k}(x)) < k \qquad \text{for every } x\in\Omega_{\varepsilon,k}\ .$$

This implies that

(2.1.5) $$g_k(x,\sigma_{\varepsilon,k}(x)) = g(x,\sigma_{\varepsilon,k}(x)) \qquad \text{for every } x\in\Omega_{\varepsilon,k}\ .$$

From (2.1.3), (2.1.4), (2.1.5) we obtain for every $x\in\Omega_{\varepsilon,k}$

(2.1.6) $$g(x,\sigma_{\varepsilon,k}(x)) + \varepsilon = g_k(x,\sigma_{\varepsilon,k}(x)) + \varepsilon < f_k(x,\sigma_{\varepsilon,k}(x)) \le f(x,\sigma_{\varepsilon,k}(x))\ .$$

From (2.1.2) and (2.1.6) it follows that

$$\int_{\Omega_{\varepsilon,k}} g(x,\sigma_{\varepsilon,k}(x))\, d\mu(x) + \varepsilon\mu(\Omega_{\varepsilon,k}) \le \int_{\Omega_{\varepsilon,k}} g(x,\sigma_{\varepsilon,k}(x))\, d\mu(x)$$

$$\int_{\Omega_{\varepsilon,k}} f(x,\sigma_{\varepsilon,k}(x))\, d\mu(x) + \varepsilon\mu(\Omega_{\varepsilon,k}) \le \int_{\Omega_{\varepsilon,k}} f(x,\sigma_{\varepsilon,k}(x))\, d\mu(x)\ .$$

By (2.1.4) and (2.1.5), taking into account hypotheses (i) and (ii), we have

(2.1.7) $$-\infty < \int_{\Omega_{\varepsilon,k}} g(x,\sigma_{\varepsilon,k}(x))\, d\mu(x) < +\infty \quad \text{or} \quad -\infty < \int_{\Omega_{\varepsilon,k}} f(x,\sigma_{\varepsilon,k}(x))\, d\mu(x) < +\infty,$$

hence, (2.1.7) implies that $\mu(\Omega_{\varepsilon,k})=0$. Setting

$$N_k = \bigcup_{\varepsilon>0} \Omega_{\varepsilon,k}$$

we have $\mu(N_k)=0$ and

$$f_k(x,s) \le g_k(x,s)$$

for every $x\in\Omega-N_k$ and for every $s\in\mathbf{R}^m$ with $|s|\le k$, and this proves (2.1.1). ■

COROLLARY 2.1.5. *Let* f *and* g *be two integrands satisfying conditions (i) and (ii) of Proposition 2.1.3 for a suitable integrand* ψ. *Then it is* $f\approx g$ *if and only if*

$$\int_B f(x,u(x))\,d\mu(x) = \int_B g(x,u(x))\,d\mu(x) \qquad \textit{for every } u\in L^\infty(\Omega;\mathbf{R}^m) \textit{ and } B\in\Im.$$

The following lemma will be useful (see also Balder [28]).

LEMMA 2.1.6. *Let* $f:\Omega\times\mathbf{R}^m\to[0,+\infty]$ *be a function such that* $f(x,\cdot)$ *is l.s.c. on* $\mathbf{R}^m$ *for* μ*-a.e.* $x\in\Omega$. *Then there exists a normal integrand* $\phi(x,s)$ *such that*

(i) $f\preceq\phi$;

(ii) *for every integrand* $g(x,s)$ *satisfying* $f\preceq g$ *it is* $\phi\preceq g$.

Proof. Without loss of generality we may assume that $f(x,\cdot)$ is l.s.c. on $\mathbf{R}^m$ for every $x\in\Omega$. Let $\mathbb{U}$ be a countable base for the open subsets of $\mathbf{R}^m$ and for every $j=(r,A)\in J=\mathbf{Q}\times\mathbb{U}$ consider the function

$$(2.1.8)\qquad f_j(s)=r\,1_A(s) \qquad \text{for every } s\in\mathbf{R}^m.$$

Since every l.s.c. function from $\mathbf{R}^m$ into $[0,+\infty]$ is the supremum of a sequence of functions of the form (2.1.8), we get

$$(2.1.9)\qquad f(x,s) = \sup\big\{f_j(s) : j\in J(x)\big\} = \sup\big\{1_{E_j}(x)f_j(s) : j\in J\big\}$$

where $J(x)$ are suitable subsets of J and $E_j=\{x\in\Omega : j\in J(x)\}$ are subsets of Ω. For every $j\in J$ let $B_j\in\Im$ such that

$$(2.1.10)\qquad E_j\subset B_j \qquad \text{and} \qquad \mu(B_j)=\mu^*(E_j)$$

where μ^* is the outer measure associated to μ defined by

$$\mu^*(E) = \inf\{\mu(B) : B\in\mathfrak{I}, B\supset E\},$$

and set

$$(2.1.11) \qquad \phi(x,s) = \sup\left\{1_{B_j}(x)f_j(s) : j\in J\right\}.$$

Then ϕ is a normal integrand such that $f \precsim \phi$, and (i) is proved.

Let now g be an integrand with $f \precsim g$; it is not restrictive to assume $f(x,s)\le g(x,s)$ for all $x\in\Omega$ and $s\in\mathbf{R}^m$. Fix $u\in L^1(\Omega;\mathbf{R}^m)$; then for every $j\in J$ the set

$$C_j = \{x\in\Omega : g(x,u(x))\ge f_j(u(x))\}$$

belongs to $\mathfrak{I}$; moreover, by (2.1.9) we get

$$E_j\subset C_j \qquad \text{for every } j\in J.$$

By (2.1.10) we obtain $\mu(B_j - C_j)=0$, so that

$$g(x,u(x)) \ge 1_{B_j}(x)f_j(u(x)) \qquad \text{for } \mu\text{-a.e. } x\in\Omega.$$

Therefore, since J is countable, by (2.1.11)

$$g(x,u(x)) \ge \phi(x,u(x)) \qquad \text{for } \mu\text{-a.e. } x\in\Omega,$$

and this concludes the proof taking into account Proposition 2.1.3. ∎

For every function $f:\Omega\to[0,+\infty]$ we define the upper and the lower integrals of f by

$$\int_\Omega^* f\,d\mu = \inf\left\{\int_\Omega g\,d\mu : g \text{ is } \mathfrak{I}\text{-measurable}, g\ge f\right\}$$

$$\int_{*\Omega} f\,d\mu = \sup\left\{\int_\Omega g\,d\mu : g \text{ is } \mathfrak{I}\text{-measurable}, g\le f\right\},$$

and for functions $f:\Omega\to[-\infty,+\infty]$

$$\int_\Omega^* f\,d\mu = \int_\Omega^* f^+\,d\mu - \int_{*\Omega} f^-\,d\mu$$

$$\int_{*\Omega} f \, d\mu = \int_{*\Omega} f^+ \, d\mu - \int_{\Omega}^{*} f^- \, d\mu$$

where f^+ and f^- denote the positive and negative parts of f, and the conventions $+\infty - \infty = +\infty$ for the upper integral and $+\infty - \infty = -\infty$ for the lower integral are adopted.

PROPOSITION 2.1.7. *Let* $f(x,s)$ *and* $\phi(x,s)$ *be as in Lemma 2.1.6. Then for every* $u \in L^1(\Omega;\mathbf{R}^m)$ *and* $B \in \Im$ *we have*

$$\int_{B}^{*} f(x,u(x)) \, d\mu(x) = \int_{B} \phi(x,u(x)) \, d\mu(x) .$$

Proof. Fix $u \in L^1(\Omega;\mathbf{R}^m)$ and $B \in \Im$, and let $g(x)$ be an $\Im$-measurable function such that

$$g(x) \geq f(x,u(x)) .$$

As in the last part of the proof of Lemma 2.1.6 we can prove that

$$g(x) \geq \phi(x,u(x)) \qquad \text{for } \mu\text{-a.e. } x \in \Omega ,$$

so that, by definition of upper integral, we have

$$\int_{\Omega}^{*} f(x,u(x)) \, d\mu(x) \geq \int_{\Omega} \phi(x,u(x)) \, d\mu(x) .$$

The opposite inequality follows from the fact that $f \leq \phi$. ■

We conclude this section with some lemmas on convex functions. If $f : \mathbf{R}^n \to]-\infty,+\infty]$ is a function, we denote by f^{**} the greatest convex l.s.c. function less than or equal to f.

LEMMA 2.1.8. *Let* $f_h:\mathbf{R}^n \to]-\infty,+\infty]$ *be an increasing sequence of l.s.c. functions. Assume there exists a function* $\theta:\mathbf{R}\to\mathbf{R}$ *such that*

(2.1.12) $$\lim_{t\to+\infty} \frac{\theta(t)}{t} = +\infty \quad \textit{and} \quad f_h(z) \geq \theta(|z|) \textit{ for every } h\in\mathbf{N},\ z\in\mathbf{R}^n.$$

Then we have

$$\sup_{h\in\mathbf{N}} f_h^{**} = \Big(\sup_{h\in\mathbf{N}} f_h\Big)^{**}.$$

Proof. The inequality $\leq$ is trivial. To prove the opposite inequality, denote by f the function $\sup\{f_h : h\in\mathbf{N}\}$ and let $g(z)=a+\langle b,z\rangle$ be an affine function less than or equal to f. By (2.1.12) we have for a suitable $k>0$

(2.1.13) $$\theta(|z|) \geq g(z) \qquad \text{on } \{z\in\mathbf{R}^n : |z|>k\}.$$

Fix a number $\varepsilon>0$; by using the lower semicontinuity of f_h and the fact that $g\leq f$ we obtain that there exists $h_0\in\mathbf{N}$ with

(2.1.14) $$f_h(z) \geq g(z) - \varepsilon \qquad \text{for every } h\geq h_0 \text{ and } |z|\leq k.$$

By (2.1.12), (2.1.13), (2.1.14) we obtain

$$f_h(z) \geq g(z) - \varepsilon \qquad \text{for every } h\geq h_0 \text{ and } z\in\mathbf{R}^n,$$

so that

$$\sup_{h\in\mathbf{N}} f_h^{**} \geq g - \varepsilon .$$

Since g and ε were arbitrary, the proof is achieved. ∎

REMARK 2.1.9. Hypothesis (2.1.12) in Lemma 2.1.8 cannot be dropped. In fact, take for instance

$$f_h(z) = \begin{cases} 1 & \text{if } |z|<h \\ 0 & \text{otherwise} \end{cases};$$

we have $\sup_{h\in\mathbf{N}} f_h^{**} = 0$ whereas $\left(\sup_{h\in\mathbf{N}} f_h\right)^{**} = 1$.

LEMMA 2.1.10. *Let* $f:\Omega\times\mathbf{R}^m\times\mathbf{R}^n\to]-\infty,+\infty]$ *be a normal integrand such that for every* $u\in L^\infty(\Omega;\mathbf{R}^m)$ *the function*

(2.1.15) $$f_u(x,z) = f(x,u(x),z)$$

is a convex integrand. Then f *is a normal-convex integrand.*

Proof. Without loss of generality we may assume that $f(x,\cdot,\cdot)$ is l.s.c. on $\mathbf{R}^m\times\mathbf{R}^n$ for every $x\in\Omega$. For every $k\in\mathbf{N}$ set

$$S_k = \{(x,s)\in\Omega\times\mathbf{R}^m : |s|\le k,\ f(x,s,\cdot) \text{ is not convex}\}$$

$$S_k(x) = \{s\in\mathbf{R}^m : (x,s)\in S_k\}$$

$$\Omega_k = \{x\in\Omega : S_k(x)\neq\emptyset\} .$$

Since f is a normal integrand we have that $(x,s)\in S_k$ if and only if $|s|\le k$ and there exist $t\in]0,1[\cap\mathbf{Q}$, $z,w\in\mathbf{Q}^n$, $h\in\mathbf{N}$ such that

$$f(x,s,tz+(1-t)w) > t\, f(x,s,z) + (1-t)\, f(x,s,w) .$$

This implies that $S_k\in\mathfrak{I}\otimes\mathbb{B}_n$. By Theorem 2.1.4 the set Ω_k belongs to $\mathfrak{I}$ and there exists an $\mathfrak{I}$-measurable function $\sigma:\Omega_k\to\mathbf{R}^m$ such that $\sigma(x)\in S_k(x)$ for every $x\in\Omega_k$. Set $\sigma(x)=0$ for $x\in\Omega-\Omega_k$. Then $\sigma\in L^\infty(\Omega;\mathbf{R}^m)$, so that by hypothesis (2.1.15) the function $f_\sigma(x,z)$ is a convex integrand and $\mu(\Omega_k)=0$. Let

$$N = \bigcup_{k\in\mathbf{N}} \Omega_k ;$$

it is $\mu(N)=0$ and, by definition of Ω_k we have that $f(x,s,\cdot)$ is convex for all $x\in\Omega-N$ and $s\in\mathbf{R}^m$. ∎

2.2. Approximation of Convex Functions

In this section we consider functions $f:\Omega\times\mathbf{R}^m\times\mathbf{R}^n\to]-\infty,+\infty]$ such that

(2.2.1) f is $\mathfrak{I}\otimes\mathbb{B}_m\otimes\mathbb{B}_n$-measurable;

(2.2.2) $f(x,\cdot,\cdot)$ is l.s.c. on $\mathbf{R}^m\times\mathbf{R}^n$ for every $x\in\Omega$;

(2.2.3) $f(x,s,\cdot)$ is convex on $\mathbf{R}^n$ for every $x\in\Omega$ and $s\in\mathbf{R}^m$.

We want to find approximations of the form

$$f(x,s,z) = \sup\{a_h(x,s) + \langle b_h(x,s),z\rangle : h\in\mathbf{N}\}$$

with $a_h(x,s)$ and $b_h(x,s)$ Carathéodory integrands. The following two selection lemmas will be useful (we refer to Castaing & Valadier [88], Theorem III.30 for the measurable selection lemma, and to Michael [225], Theorem 3.1 for the continuous selection lemma).

LEMMA 2.2.1. (measurable selections). *Let* Y *be a complete separable metric space and let* $T:\Omega\to\wp(Y)$ *be a multimapping. Assume that*

(i) *for every* $x\in\Omega$ *the set* $T(x)$ *is closed and non empty;*

(ii) T *is* $\mathfrak{I}\otimes\mathbb{B}(Y)$*-measurable (*$\mathbb{B}(Y)$ *denotes the Borel* σ*-algebra of* Y*), in the sense that its graph* $\{(x,y)\in\Omega\times Y : y\in T(x)\}$ *belongs to* $\mathfrak{I}\otimes\mathbb{B}(Y)$.

Then, there exists a sequence of measurable functions $\sigma_h:(\Omega,\mathfrak{I})\to(Y,\mathbb{B}(Y))$ *such that for every* $x\in\Omega$ *the set* $\{\sigma_h(x) : h\in\mathbf{N}\}$ *is dense in* $T(x)$.

LEMMA 2.2.2. (continuous selections). *Let* Y *be a metrizable space and let* $T:Y\to\wp(\mathbf{R}^k)$ *be a multimapping. Assume that*

(i) *for every* $y\in Y$ *the set* $T(y)$ *is closed, convex, and non empty;*

(ii) T *is lower semicontinuous, that is for every open set* $U\subset\mathbf{R}^k$ *the set*

$$T^{-}(U)=\{y\in Y : T(y)\cap U\neq\emptyset\}$$

is open in Y.

Then, for every $y_0\in Y$ *and every* $v_0\in T(y_0)$ *there exists a continuous function* $\sigma:Y\to \mathbf{R}^k$ *such that*

$$\begin{cases} \sigma(y)\in T(y) & \textit{for every } y\in Y \\ \sigma(y_0)=v_0 \ . \end{cases}$$

When the function $f(x,s,z)$ does not depend on the x variable, the following result holds.

LEMMA 2.2.3. *Let* $f:\mathbf{R}^m\times\mathbf{R}^n\to]-\infty,+\infty]$ *be a l.s.c. function with* $f(s,\cdot)$ *convex for all* $s\in\mathbf{R}^m$. *Assume one of the following conditions is satisfied:*

(i) *there exists a continuous function* $z_0:\mathbf{R}^m\to\mathbf{R}^n$ *such that the function* $s\to f(s,z_0(s))$ *is continuous and finite;*

(ii) *there exists a function* $\theta:\mathbf{R}\to\mathbf{R}$ *such that*

$$\lim_{t\to+\infty}\frac{\theta(t)}{t}=+\infty \qquad \textit{and} \qquad f(s,z)\geq\theta(|z|) \textit{ for every } s\in\mathbf{R}^m,\ z\in\mathbf{R}^n.$$

Then, there exist two sequences of continuous functions $a_h:\mathbf{R}^m\to\mathbf{R}$ *and* $b_h:\mathbf{R}^m\to\mathbf{R}^n$ *such that*

(2.2.4) $\quad f(s,z) = \sup\{a_h(s)+\langle b_h(s),z\rangle : h\in\mathbf{N}\}\quad$ *for every* $s\in\mathbf{R}^m$, $z\in\mathbf{R}^n$.

Proof. Assume first condition (i) is satisfied. By considering the function

$$f(s,z+z_0(s)) - f(s,z_0(s))$$

we may assume $f(s,0)=0$. Let $T:\mathbf{R}^m\to\wp(\mathbf{R}\times\mathbf{R}^n)$ be defined by

$$T(s) = \{(a,b)\in\mathbf{R}\times\mathbf{R}^n : f(s,z)\geq a+\langle b,z\rangle \text{ for every } z\in\mathbf{R}^n\}.$$

For every $s\in\mathbf{R}^m$ the set $T(s)$ is closed, convex, and non empty; we prove that the multimapping T is lower semicontinuous. Fix an open set $U\subset\mathbf{R}\times\mathbf{R}^n$; we have to show that the set $T^-(U)=\{s\in\mathbf{R}^m : T(s)\cap U\neq\emptyset\}$ is open in $\mathbf{R}^m$. If $T^-(U)$ is not empty, fix $s_0\in T^-(U)$ and $(a_0,b_0)\in T(s_0)\cap U$; since U is open, we may assume that

(2.2.5) $$\{(a,b) : |a-a_0| + |b-b_0| \le 2\varepsilon\} \subset U$$

for a suitable $\varepsilon>0$. We claim there exists a neighbourhood A of s_0 in $\mathbf{R}^m$ such that

(2.2.6) $$f(s,z) \ge a_0 + \langle b_0,z\rangle - \varepsilon(1+|z|) \qquad \text{for every } s\in A,\ z\in\mathbf{R}^n .$$

In fact, if (2.2.6) were false, it would be possible to find a sequence (s_h) converging to s_0 in $\mathbf{R}^m$, and a sequence (z_h) in $\mathbf{R}^n$ such that

$$f(s_h,z_h) < a_0 + \langle b_0,z_h\rangle - \varepsilon(1+|z_h|) \qquad \text{for every } h\in\mathbf{N}.$$

If (z_h) is bounded, we may assume that $z_h\to z_0$ in $\mathbf{R}^n$, and the lower semicontinuity of f yields

$$f(s_0,z_0) \le \liminf_{h\to\infty} f(s_h,z_h) \le a_0 + \langle b_0,z_0\rangle - \varepsilon(1+|z_0|)$$

which contradicts the fact that $(a_0,b_0)\in T(s_0)$. If (z_h) is not bounded, we may assume

$$|z_h| \to +\infty \qquad \text{and} \qquad w_h = \frac{z_h}{|z_h|} \to w_0 \ \text{ in } \mathbf{R}^n .$$

Then, for every $t>0$

$$f(s_h,tw_h) \le \frac{t}{|z_h|} f(s_h,z_h) \le \frac{t}{|z_h|}\big(a_0 + \langle b_0,z_h\rangle - \varepsilon(1+|z_h|)\big) ,$$

and, letting $h\to\infty$,

$$f(s_0,tw_0) \le \langle b_0,tw_0\rangle - \varepsilon t ,$$

which contradicts again (taking t sufficiently large) the fact that $(a_0,b_0)\in T(s_0)$. Therefore (2.2.6) is proved.

Let now $s\in A$ and let $(a,b)\in\mathbf{R}\times\mathbf{R}^n$ be such that $|a|\le\varepsilon$, $|b|\le\varepsilon$ and

$$f(s,z) - a_0 - \langle b_0,z\rangle \geq a + \langle b,z\rangle \geq -\varepsilon(1+|z|) \qquad \text{for all } z\in \mathbf{R}^n.$$

We have

$$f(s,z) \geq (a+a_0) - \langle b+b_0,z\rangle \qquad \text{for all } z\in \mathbf{R}^n,$$

so that $(a+a_0,b+b_0)\in T(s)$. Moreover, it is $|a|+|b|\leq 2\varepsilon$; hence by (2.2.5) we have $(a+a_0,b+b_0)\in U$, and this proves that the multimapping T is lower semicontinuous. By using the continuous selection Lemma 2.2.2 we find two families $\{a_i(s)\}_{i\in I}$ and $\{b_i(s)\}_{i\in I}$ of continuous functions such that

$$f(s,z) = \sup \{a_i(s) + \langle b_i(s),z\rangle \ : \ i\in I\} .$$

Since f is l.s.c., by the Lindelöf covering lemma the set I can be taken countable, so that the proof is achieved under hypothesis (i).

Assume now condition (ii) is satisfied. For every $h\in \mathbf{N}$ define

$$f_h(s,z) = \inf \{f(t,z) + h|s-t| \ : \ t\in \mathbf{R}^m\} ;$$

the functions f_h satisfy the following properties:

(2.2.7) $\sup\{f_h : h\in \mathbf{N}\} = f$;

(2.2.8) $f_h(s,z) \leq f_h(t,z) + h|s-t|$ for every $h\in \mathbf{N}$, $s,t\in \mathbf{R}^m$, $z\in \mathbf{R}^n$;

(2.2.9) for every $h\in \mathbf{N}$ and $s\in \mathbf{R}^m$ the function $f_h(s,\cdot)$ is l.s.c. on $\mathbf{R}^n$.

Properties (2.2.7) and (2.2.8) follow from Proposition 1.3.7. In order to prove (2.2.9) fix $h\in \mathbf{N}$, $s\in \mathbf{R}^m$, $z\in \mathbf{R}^n$, and $z_k\to z$ in $\mathbf{R}^n$ with

$$\lim_{k\to\infty} f_h(s,z_k) < +\infty . \tag{2.2.10}$$

By the definition of f_h there exists a sequence (s_k) in $\mathbf{R}^m$ with

$$f(s_k,z_k) + h|s-s_k| \leq f_h(s,z_k) + 2^{-k} \qquad \text{for every } k\in \mathbf{N} .$$

By (2.2.10) the sequence (s_k) is bounded, so we may assume $s_k\to t$ for a suitable $t\in \mathbf{R}^m$. By the lower semicontinuity of f we have

$$f_h(s,z) \leq f(t,z) + h|s-t| \leq$$

$$\leq \liminf_{k\to\infty} \left[f(s_k,z_k) + h|s-s_k|\right] \leq \liminf_{k\to\infty} f_h(s,z_k) ,$$

so that (2.2.9) is proved.

By Lemma 2.1.8 we obtain for every $s\in\mathbf{R}^m$

$$f(s,\cdot) = \sup\{f_h^{**}(s,\cdot) : h\in\mathbf{N}\} . \tag{2.2.11}$$

By using part (i) of the lemma, for every $h\in\mathbf{N}$ we have

$$f_h^{**}(s,z) = \sup\left\{a_{h,k}(s) + \langle b_{h,k}(s),z\rangle \ : \ k\in\mathbf{N}\right\}$$

for suitable continuous functions $a_{h,k}$ and $b_{h,k}$; therefore, by (2.2.11) formula (2.2.4) follows. ∎

We consider now the general case of a function $f(x,s,z)$.

THEOREM 2.2.4. *Let* $f:\Omega\times\mathbf{R}^m\times\mathbf{R}^n\to]-\infty,+\infty]$ *be a function satisfying assumptions (2.2.1), (2.2.2), (2.2.3). Assume further that for every* $x\in\Omega$ *the function* $f(x,\cdot,\cdot)$ *satisfies one of conditions (i), (ii) of Lemma 2.2.3. Then, there exist two sequences* $a_h:\Omega\times\mathbf{R}^m\to\mathbf{R}$ *and* $b_h:\Omega\times\mathbf{R}^m\to\mathbf{R}^n$ *of Carathéodory integrands such that*

$$f(x,s,z) = \sup\{a_h(x,s) + \langle b_h(x,s),z\rangle : h\in\mathbf{N}\} \quad \textit{for every } x\in\Omega,\ s\in\mathbf{R}^m,\ z\in\mathbf{R}^n.$$

Proof. Let Y be that space of continuous functions from $\mathbf{R}^m$ into $\mathbf{R}\times\mathbf{R}^n$ endowed with the usual distance inducing the topology of uniform convergence on compact sets

$$d(f,g) = \sum_{h\in\mathbf{N}} 2^{-h}\operatorname{arctg}\left(\|f-g\|_{L^\infty(K_h)}\right)$$

where (K_h) is an increasing sequence of compact subsets of $\mathbf{R}^m$ whose union is $\mathbf{R}^m$. It is well known that the metric space (Y,d) is complete and separable. For every $y\in Y$ we denote by y^1 and y^2 its components with values in $\mathbf{R}$ and $\mathbf{R}^n$ respectively, and we define

$$w_y(s,z) = y^1(s) + \langle y^2(s),z\rangle .$$

Let A be an open subset of $\mathbf{R}^m \times \mathbf{R}^n$; set

$$w_A(y) = \inf\{w_y(s,z) : (s,z)\in A\}$$

$$f_A(x) = \inf\{f(x,s,z) : (s,z)\in A\}$$

$$T(x) = \{y\in Y : f(x,s,z)\geq w_y(s,z) \text{ for every } s\in \mathbf{R}^m,\ z\in \mathbf{R}^n\} .$$

It is immediately seen that w_A is upper semicontinuous in Y; moreover, f_A is measurable on Ω. In fact, for every $c\in\mathbf{R}$, setting $A_c=\{x\in\Omega : f_A(x)<c\}$, we have

$$A_c = \{x\in\Omega : f(x,s,z)<c \text{ for some } (s,z)\in A\} =$$

$$= \Pi_1\{(x,s,z)\in \Omega\times\mathbf{R}^m\times\mathbf{R}^n : f(x,s,z)<c,\ (s,z)\in A\}$$

where $\Pi_1 : \Omega\times\mathbf{R}^m\times\mathbf{R}^n \to \Omega$ denotes the projection on Ω. Then, by the measurable projection theorem (see Castaing & Valadier [88], Theorem III.23) the set A_c belongs to $\mathfrak{I}$.

Finally, the multimapping T is $\mathfrak{I}\otimes\mathbb{B}(Y)$-measurable. In fact, if $\mathbb{U}$ is a countable base for the open subsets of $\mathbf{R}^m\times\mathbf{R}^n$, we have

$$\Big\{(x,y)\in\Omega\times Y : y\in T(x)\Big\} = \bigcap_{A\in\mathbb{U}} \Big\{(x,y)\in\Omega\times Y : f_A(x)\geq w_A(y)\Big\}$$

so that the graph of T belongs to $\mathfrak{I}\otimes\mathbb{B}(Y)$.

By the measurable selection Lemma 2.2.1 we find a sequence of measurable functions $y_h : \Omega\to Y$ such that

$$T(x) = \overline{\{y_h(x) : h\in\mathbf{N}\}} \qquad \text{for every } x\in\Omega . \tag{2.2.12}$$

Define now for every $h\in\mathbf{N}$, $x\in\Omega$, $s\in\mathbf{R}^m$

$$a_h(x,s) = y_h^1(x)(s) \qquad \text{and} \qquad b_h(x,s) = y_h^2(x)(s) .$$

Then a_h and b_h are Carathéodory functions; moreover, by (2.2.12) and by Lemma 2.2.3 we have

$$f(x,s,z) = \sup \{a_h(x,s) + \langle b_h(x,s),z\rangle : h\in \mathbf{N}\}$$

for every $x\in\Omega$, $s\in\mathbf{R}^m$, $z\in\mathbf{R}^n$. ∎

REMARK 2.2.5. Let f, a_h, b_h as in the statement of Theorem 2.2.4, with $f\geq 0$. For every $h,k\in\mathbf{N}$, $x\in\Omega$, $s\in\mathbf{R}^m$ define

$$d_{h,k}(x,s) = \frac{k}{k\vee(|a_h(x,s)|+|b_h(x,s)|)}$$

$$a_{h,k}(x,s) = d_{h,k}(x,s)\, a_h(x,s)$$

$$b_{h,k}(x,s) = d_{h,k}(x,s)\, b_h(x,s) .$$

It is easy to see that the following approximation formula for f holds:

$$f(x,s,z) = \sup \left\{ \left[a_{h,k}(x,s) + \langle b_{h,k}(x,s),z\rangle\right]^+ : h\in\mathbf{N}, k\in\mathbf{N} \right\} .$$

Note that the functions $a_{h,k}$ and $b_{h,k}$ are bounded Carathéodory integrands.

REMARK 2.2.6. Theorem 2.2.3 (hence Theorem 2.2.4 and Remark 2.2.5) holds with the same proof also for sequentially l.s.c. functions $f:X\times Y\to]-\infty,+\infty]$ where X is a general metric space and Y is a reflexive separable Banach space endowed with its weak topology. Moreover, if hypothesis (i) is substituted by

(i') *there exists* $z_0\in L^\infty(X;Y)$ *such that* $f(s,z_0(s))$ *is bounded,*

then it is possible to prove (see Olech [239], Theorem 1 and Corollary 3) that the approximation formula (2.2.4) holds for suitable measurable functions a_h and continuous functions b_h.

We conclude this section by an application of the measurable selection Lemma 2.2.1 to minimum points of normal integrands.

PROPOSITION 2.2.7. *Let* $f:\Omega\times\mathbf{R}^m\to[0,+\infty]$ *be a normal integrand such that*

(2.2.13) *for* μ*-a.e.* $x\in\Omega$ *the function* $f(x,\cdot)$ *attains its minimum on* $\mathbf{R}^m$.

Then, there exists a measurable mapping $u_0:\Omega\to\mathbf{R}^m$ *such that*

$$f(x,u_0(x)) = \min\{f(x,s) : s\in\mathbf{R}^m\} \qquad \text{for } \mu\text{-a.e. } x\in\Omega.$$

Proof. It is not restrictive to assume that for every $x\in\Omega$ the function $f(x,\cdot)$ is l.s.c. and attains its minimum on $\mathbf{R}^m$. Denote by $T:\Omega\to\wp(\mathbf{R}^m)$ the multimapping

$$T(x) = \{s\in\mathbf{R}^m : s \text{ is a minimum point of } f(x,\cdot)\} ;$$

the lower semicontinuity of $f(x,\cdot)$ and (2.2.13) imply that for every $x\in\Omega$ the set $T(x)$ is closed and non empty. Therefore, by the measurable selection Lemma 2.2.1, it will be enough to check that T is $\mathfrak{I}\otimes\mathbb{B}_m$-measurable, that is

(2.2.14) $\{(x,s)\in\Omega\times\mathbf{R}^m : s\in T(x)\}\in\mathfrak{I}\otimes\mathbb{B}_m$.

For every $k\in\mathbf{N}$ denote by $f_k(x,s)$ the Moreau-Yosida transform of f:

$$f_k(x,s) = \inf\{f(x,t) + k|s-t| : t\in\mathbf{R}^m\} .$$

It is clear that for every $s\in T(x)$ we have

(2.2.15) $f_k(x,t) \geq f(x,s)$ for every $t\in\mathbf{R}^m$ and every $k\in\mathbf{N}$.

On the other hand, if (2.2.15) holds for some $s\in\mathbf{R}^m$, by Proposition 1.3.7 we have $s\in T(x)$, so that (2.2.14) becomes

$$\bigcap_{k\in\mathbf{N}}\left\{(x,s)\in\Omega\times\mathbf{R}^m : f(x,s)\leq f_k(x,t) \quad \forall t\in\mathbf{R}^m\right\}\in\mathfrak{I}\otimes\mathbb{B}_m .$$

Since by Proposition 1.3.7 every $f_k(x,\cdot)$ is a continuous function, it will be enough to prove that for every $k\in\mathbf{N}$ and every $t\in\mathbf{Q}$

$$\{(x,s)\in\Omega\times\mathbf{R}^m : f(x,s)\le f_k(x,t)\}\in\Im\otimes\mathbb{B}_m,$$

which follows from the fact that f is $\Im\otimes\mathbb{B}_m$-measurable. ■

2.3. Two Lower Semicontinuity Results

In this section we shall present two lower semicontinuity results (Theorems 2.3.1 and 2.3.6) for functionals of the form

$$F(u,v) = \int_\Omega f(x,u(x),v(x))\, d\mu(x) .$$

As far as we know the first proof of Theorem 2.3.1 has been given by De Giorgi in [126] for Carathéodory integrands $f(x,s,z)\ge 0$; the more general version we present here is due to Ioffe (see [183]), even if our proof is slightly different from the original one. The first lower semicontinuity result is the following.

THEOREM 2.3.1. *Let* $p,q\in[1,+\infty]$, *and let* $f:\Omega\times\mathbf{R}^m\times\mathbf{R}^n\to]-\infty,+\infty]$ *be a function satisfying conditions (2.2.1),(2.2.2),(2.2.3). Assume further the following inf-compactness property holds:*

for every sequence (u_h) *converging in* $L^p(\Omega;\mathbf{R}^m)$ *and every sequence* (v_h) *weakly converging in* $L^q(\Omega;\mathbf{R}^n)$ *(weakly* if* $q=+\infty$*) such that for a suitable constant* $c>0$

$$\int_\Omega f^+(x,u_h(x),v_h(x))\, d\mu(x) \le c + \int_\Omega f^-(x,u_h(x),v_h(x))\, d\mu(x) \qquad \forall h\in\mathbf{N},$$

the sequence $f^-(x,u_h(x),v_h(x))$ *is weakly compact in* $L^1(\Omega)$.

Then, the functional

$$F(u,v) = \int_\Omega f(x,u,v)\, d\mu(x) \qquad u\in L^p(\Omega;\mathbf{R}^m),\ v\in L^q(\Omega;\mathbf{R}^n)$$

is well-defined, takes its values in $]-\infty,+\infty]$ *and is sequentially l.s.c. on* $L^p(\Omega;\mathbf{R}^m)\times L^q(\Omega;\mathbf{R}^n)$ *with respect to the strong topology of* $L^p(\Omega;\mathbf{R}^m)$ *and the weak topology of* $L^q(\Omega;\mathbf{R}^n)$ *(weak* if* $q=+\infty$*).*

For the proof of Theorem 2.3.1 the following lemma will be useful.

LEMMA 2.3.2. *Let* g_h *and* g *be* μ*-measurable functions from* Ω *into* $]-\infty,+\infty]$ *such that* $g=\sup\{g_h : h\in\mathbf{N}\}$ *and* $g_h\geq\gamma$ *for a suitable* $\gamma\in L^1(\Omega)$. *Then*

$$\int_\Omega g\, d\mu = \sup\Big\{\sum_{i\in I}\int_{B_i} g_i\, d\mu\Big\}$$

where the supremum is taken over all finite partitions of Ω *by pairwise disjoint* $B_i\in\mathfrak{I}$. *In particular we have*

$$\int_\Omega g^+\, d\mu = \sup\Big\{\int_B g\, d\mu \ :\ B\in\mathfrak{I}\Big\}\ .$$

for every $g\in L^1(\Omega)$.

Proof. The inequality $\geq$ is trivial. To prove the opposite inequality, set $f_h=g_1\vee\ldots\vee g_h$; then $f_h\uparrow g$, and the Beppo Levi monotone convergence theorem implies

$$\int_\Omega g\, d\mu = \sup\Big\{\int_\Omega f_h\, d\mu \ :\ h\in\mathbf{N}\Big\}\ .$$

Now, there exist pairwise disjoint sets $B_1,\ldots,B_h\in\mathfrak{I}$ such that

$$\Omega = \bigcup_{i=1}^{h} B_i \qquad \text{and} \qquad f_h = g_i \text{ on } B_i ,$$

so that

$$\int_\Omega f_h \, d\mu = \sum_{i=1}^{h} \int_{B_i} g_i \, d\mu ,$$

and the lemma is proved. ■

Proof of Theorem 2.3.1. The fact that the functional F is well-defined and takes its values in $]-\infty,+\infty]$ follows immediately from the inf-compactness condition. In fact, take $u \in L^p(\Omega;\mathbf{R}^m)$ and $v \in L^q(\Omega;\mathbf{R}^n)$; if for a suitable constant $c>0$

$$(2.3.1) \qquad \int_\Omega f^+(x,u(x),v(x)) \, d\mu(x) \le c + \int_\Omega f^-(x,u(x),v(x)) \, d\mu(x) ,$$

then from the inf-compactness condition we have $f^-(x,u(x),v(x)) \in L^1(\Omega)$, so that by using (2.3.1) $f(x,u(x),v(x)) \in L^1(\Omega)$. If (2.3.1) is not satisfied, we have

$$\int_\Omega f^-(x,u(x),v(x)) \, d\mu(x) < +\infty \qquad \text{and} \qquad \int_\Omega f^+(x,u(x),v(x)) \, d\mu(x) = +\infty ,$$

so that

$$\int_\Omega f(x,u(x),v(x)) \, d\mu(x) = +\infty .$$

The sequential lower semicontinuity will be proved in several steps.

Step 1. We prove the result under the additional condition:

there exists a function $\theta:\mathbf{R}\to\mathbf{R}$ *such that*

$$\lim_{t\to+\infty} \frac{\theta(t)}{t} = +\infty \qquad \textit{and} \qquad f(x,s,z) \ge \theta(|z|) .$$

In this case, by Theorem 2.2.4 and Remark 2.2.5 we have

$$f(x,s,z) = \sup\left\{[a_h(x,s) + \langle b_h(x,s),z\rangle]^+ \ :\ h\in\mathbf{N}\right\}$$

for suitable sequences of bounded Carathéodory integrands a_h and b_h. By Lemma 2.3.2 it will be enough to prove that for every $h\in\mathbf{N}$ and $B\in\mathfrak{I}$ the functional

$$\int_B a_h(x,u)+\langle b_h(x,u),v\rangle\, d\mu$$

is sequentially l.s.c. with respect to the strong topology of $L^p(\Omega;\mathbf{R}^m)$ and the weak topology of $L^q(\Omega;\mathbf{R}^n)$ (weak* if $q=+\infty$), but this is trivial because a_h and b_h are bounded Carathéodory integrands.

Step 2. We prove the result under the additional condition:

f *is bounded from below.*

We may assume for instance that f is positive. Let $u_h\to u$ in $L^p(\Omega;\mathbf{R}^m)$ and $v_h\to v$ weakly in $L^q(\Omega;\mathbf{R}^n)$ (weakly* if $q=+\infty$); by Theorem 1.2.8 there exists a function $\theta:\mathbf{R}\to\mathbf{R}$ such that

$$\lim_{t\to+\infty}\frac{\theta(t)}{t} = +\infty \qquad \text{and} \qquad \int_\Omega \theta(|v_h|)\, d\mu \le 1 \quad \text{for every } h\in\mathbf{N}.$$

Setting for every $\varepsilon>0$

$$f_\varepsilon(x,s,z) = f(x,s,z) + \varepsilon\theta(|z|)$$

we have by Step 1

$$\int_\Omega f(x,u,v)\, d\mu \le \int_\Omega f_\varepsilon(x,u,v)\, d\mu \le$$

$$\le \liminf_{h\to+\infty}\int_\Omega f_\varepsilon(x,u_h,v_h)\, d\mu \le \varepsilon + \liminf_{h\to+\infty}\int_\Omega f(x,u_h,v_h)\, d\mu\ ,$$

and, since ε was arbitrary, the proof of Step 2 is achieved.

Step 3. We prove now the result in the general case. Let $u_h\to u$ in $L^p(\Omega;\mathbf{R}^m)$ and $v_h\to v$ weakly in $L^q(\Omega;\mathbf{R}^n)$ (weakly* if $q=+\infty$), and set for every $h,k\in\mathbf{N}$

$$f_k(x,s,z) = f(x,s,z)\vee(-k)$$

$$F_k(u,v) = \int_\Omega f_k(x,u,v)\,d\mu$$

$$g_h(x) = f^-(x,u_h(x),v_h(x))$$

$$A_{h,k} = \{x\in\Omega : g_h>k\}.$$

We may assume that the sequence $F(u_h,v_h)$ tends to some finite limit as $h\to+\infty$, so that, by the inf-compactness property, the sequence $\{g_h\}$ turns out to be weakly compact in $L^1(\Omega)$. By Step 2 we have for every $k\in\mathbf{N}$

$$F(u,v) \le F_k(u,v) \le \liminf_{h\to+\infty} F_k(u_h,v_h) =$$

$$= \liminf_{h\to+\infty} \Big[F(u_h,v_h) + \int_{A_{h,k}}(g_h-k)\,d\mu\Big] \le$$

$$\le \liminf_{h\to+\infty} F(u_h,v_h) + \limsup_{h\to+\infty} \int_{A_{h,k}} g_h\,d\mu \quad ,$$

and, since Theorem 1.2.8 implies that g_h are uniformly-integrable, the proof of the theorem follows by taking $k\to+\infty$. ■

With some slight modifications in the proof, Theorem 2.3.1 can be generalized to functionals of the form

$$F(u,v) = \int_\Omega^* f(x,u(x),v(x))\,d\mu(x)$$

where $f(x,s,z)$ is not necessarily measurable. More precisely, the following result holds.

THEOREM 2.3.3. *Let* $p,q\in[1,+\infty]$, *and let* $f:\Omega\times\mathbf{R}^m\times\mathbf{R}^n\to[-\infty,+\infty]$ *be a function* *satisfying conditions (2.2.2) and (2.2.3) (remark that no measurability assumptions are made). For every* $u \in L^p(\Omega;\mathbf{R}^m)$ *and* $v \in L^q(\Omega;\mathbf{R}^n)$ *define*

$$F(u,v) = \int_{\Omega}^{*} f(x,u(x),v(x))\, d\mu(x)$$

and assume the following inf-compactness property holds:

for every sequence (u_h) *converging in* $L^p(\Omega;\mathbf{R}^m)$ *and every sequence* (v_h) *weakly converging in* $L^q(\Omega;\mathbf{R}^n)$ *(weakly* if* $q=+\infty$*) with*

$$\sup\{F(u_h,v_h) : h \in \mathbf{N}\} < +\infty ,$$

the sequence $g_h(x)=f^-(x,u_h(x),v_h(x))$ *is equi-integrable in the sense that*

$$\sup_{h\in\mathbf{N}} \int_{\Omega}^{*} g_h\, d\mu < +\infty \qquad \text{and} \qquad \lim_{\mu(A)\to 0} \Big\{ \sup_{h\in\mathbf{N}} \int_{A}^{*} g_h\, d\mu \Big\} = 0 .$$

Then, the functional F *takes its values in* $]-\infty,+\infty]$ *and is sequentially l.s.c. on* $L^p(\Omega;\mathbf{R}^m)\times L^q(\Omega;\mathbf{R}^n)$ *with respect to the strong topology of* $L^p(\Omega;\mathbf{R}^m)$ *and the weak topology of* $L^q(\Omega;\mathbf{R}^n)$ *(weak* if* $q=+\infty$*).*

In order to study necessary conditions for the lower semicontinuity we need some properties of non-atomic measures.

REMARK 2.3.4. The Lebesgue measure on $\mathbf{R}^n$ is non-atomic. Moreover (see Rudin [254], page 114), if $\mu_1,\ldots,\mu_k$ are positive, finite and non-atomic measures on $(\Omega,\mathfrak{I})$, the set $\{(\mu_1(B),\ldots,\mu_k(B)) : B\in\mathfrak{I}\}$ is a compact convex subset of $\mathbf{R}^k$.

PROPOSITION 2.3.5. *For every positive finite measure* μ *on* $(\Omega,\mathfrak{I})$ *the following conditions are equivalent:*

(i) μ *is non-atomic;*

(ii) *for every* $t \in [0,1]$ *there exists a sequence* $(A(h))_{h\in\mathbf{N}}$ *in* $\mathfrak{I}$ *such that* $1_{A(h)} \to t$ *in the weak* topology of* $L^\infty(\Omega,\mu)$.

Proof. We prove (ii)⇒(i). Let A be a non-negligible element of $\mathfrak{I}$; then by (ii) there exists a sequence $\{A(h)\}_{h\in\mathbf{N}}$ in $\mathfrak{I}$ such that

$$\lim_{h\to+\infty} \mu(A\cap A(h)) = \frac{1}{2}\mu(A) ,$$

and this proves that μ is non-atomic.

We prove (i)⇒(ii). Denote by I the interval $[0,1]$ and by ν the Lebesgue measure on I. We want to construct a measurable function $f:\Omega\to I$ such that

(2.3.2) $$\mu(\Omega)\,\nu(B) = \mu(f^{-1}(B))$$

for every Borel subset B of I. For every $k\in\mathbf{N}$ let

$$\Lambda_k = \left\{ (i_1,\dots,i_k) : i_j \in \{0,1\} \text{ for } j=1,\dots,k \right\} \quad \text{and} \quad \Lambda = \bigcup_{k\in\mathbf{N}} \Lambda_k .$$

Since μ is non-atomic, by Remark 2.3.4 every non-negligible set can be divided in two parts having the same measure; therefore, for every $\lambda\in\Lambda$ we can construct by induction a set Ω_λ in the following way:

$$\Omega_0 \cup \Omega_1 = \Omega \quad \text{with} \quad \mu(\Omega_0) = \mu(\Omega_1) = \frac{1}{2}\mu(\Omega)$$

$$\Omega_{i_1,\dots,i_k,0} \cup \Omega_{i_1,\dots,i_k,1} = \Omega_{i_1,\dots,i_k}$$

$$\text{with} \quad \mu(\Omega_{i_1,\dots,i_k,0}) = \mu(\Omega_{i_1,\dots,i_k,1}) = \frac{1}{2}\mu(\Omega_{i_1,\dots,i_k}) .$$

Finally, define $f_k:\Omega\to I$ by

$$f_k(x) = \sum_{j=1}^{k} 2^{-j}\, i_j \qquad \text{if } x\in\Omega_{i_1,\dots,i_k} .$$

The functions f_k are measurable, increasing with respect to k, and converge uniformly to a measurable strictly increasing function f. Moreover, it is easy to check that formula (2.3.2) holds.

Let now $t\in[0,1]$; there exists a sequence $\{I(h)\}_{h\in\mathbf{N}}$ of Borel subsets of I (take for instance $I(h)=\{x\in I : hx-[hx]<t\}$) such that $1_{I(h)}\to t$ weakly* in $L^\infty(I)$. Take $A(h)=f^{-1}(I(h))$, let $g\in L^1(\Omega)$, and set

$$G(y) = g(f^{-1}(y)) \qquad \text{for every } y\in I.$$

By the properties of measures (see for instance Dellacherie & Meyer [138]) we have

$$\lim_{h\to+\infty}\int_{A(h)} g(x)\,d\mu(x) = \lim_{h\to+\infty}\mu(\Omega)\int_{I(h)} G(y)\,d\nu(y) =$$

$$= t\,\mu(\Omega)\int_I G(y)\,d\nu(y) = t\int_\Omega g(x)\,d\mu(x) \;,$$

which proves that $1_{A(h)}\to t$ weakly* in $L^\infty(\Omega)$. ∎

We can now pass to the second lower semicontinuity result.

THEOREM 2.3.6. *Let* $f:\Omega\times\mathbf{R}^m\times\mathbf{R}^n\to]-\infty,+\infty]$ *be a normal integrand such that*

(2.3.3) $$f(x,s,z) \geq -\psi(x,|s|,|z|) \quad \textit{for every } (x,s,z)\in\Omega\times\mathbf{R}^m\times\mathbf{R}^n$$

for a suitable integrand $\psi(x,t,\tau)$ *defined on* $\Omega\times[0,+\infty[\times[0,+\infty[$ *increasing in* t *and* τ *and* μ*-integrable in* x. *Consider the functional*

$$F(u,v) = \int_\Omega f(x,u,v)\,d\mu$$

defined for every $u\in L^\infty(\Omega;\mathbf{R}^m)$ *and* $v\in L^\infty(\Omega;\mathbf{R}^n)$, *and assume that:*

(i) *for every* $u\in L^\infty(\Omega;\mathbf{R}^m)$ *the functional* $F(u,\cdot)$ *is sequentially weakly* l.s.c. on* $L^\infty(\Omega;\mathbf{R}^n)$;

(ii) *for every* $u \in L^\infty(\Omega;\mathbf{R}^m)$ *there exists* $v \in L^\infty(\Omega;\mathbf{R}^n)$ *such that* $F(u,v)<+\infty$.

Then the function $f(x,s,z)$ *is a normal-convex integrand.*

Proof. Let $u \in L^\infty(\Omega;\mathbf{R}^m)$ be fixed; by assumption (ii) there exists $v_0 \in L^\infty(\Omega;\mathbf{R}^n)$ such that $f(x,u(x),v_0(x)) \in L^1(\Omega)$. Let $t \in [0,1]$, $B \in \mathfrak{I}$, and $v,w \in L^\infty(\Omega;\mathbf{R}^n)$; we will show that

$$\int_B f(x,u,tv+(1-t)w)\,d\mu \le \int_B \big[t\,f(x,u,v) + (1-t)\,f(x,u,w)\big]\,d\mu\,. \tag{2.3.4}$$

It is enough to consider only the case when $f(x,u(x),v(x))$ and $f(x,u(x),w(x))$ are in $L^1(B)$. By Proposition 2.3.5 there exists a sequence $\{A(h)\}_{h \in \mathbf{N}}$ in $\mathfrak{I}$ such that $1_{A(h)} \to t$ in the weak* topology of $L^\infty(\Omega;\mu)$. Then, by hypothesis (i) we have

$$\int_B f(x,u,tv+(1-t)w)\,d\mu =$$

$$= F(u,tv\cdot 1_B+(1-t)w\cdot 1_B+v_0\cdot 1_{\Omega-B}) - \int_{\Omega-B} f(x,u,v_0)\,d\mu \le$$

$$\le \liminf_{h\to+\infty} F(u,v\cdot 1_{B\cap A(h)}+w\cdot 1_{B-A(h)}+v_0\cdot 1_{\Omega-B}) - \int_{\Omega-B} f(x,u,v_0)\,d\mu =$$

$$= \liminf_{h\to+\infty} \int_{B\cap A(h)} f(x,u,v)\,d\mu + \int_{B-A(h)} f(x,u,w)\,d\mu =$$

$$= \int_B \big[t\,f(x,u,v) + (1-t)\,f(x,u,w)\big]\,d\mu\,,$$

so that (2.3.4) is proved. Hence, by Proposition 2.1.3, there exists a μ-negligible set $N \subset \Omega$ such that

$$f(x,u(x),tz_1+(1-t)z_2) \le t\,f(x,u(x),z_1) + (1-t)\,f(x,u(x),z_2) \tag{2.3.5}$$

for every $x \in \Omega - N$, $z_1,z_2 \in \mathbf{R}^n$, $t \in [0,1]\cap\mathbf{Q}$. Taking into account that f is a normal integrand, that is the function $f(x,\cdot,\cdot)$ is l.s.c. for μ-a.e. $x \in \Omega$, we get that there exists a μ-negligible set $N'\subset\Omega$ such that (2.3.5) holds for every $x\in\Omega-N'$, $z_1,z_2\in\mathbf{R}^n$, and $t\in[0,1]$. Finally, by Lemma 2.1.10, this implies that f is a normal-convex integrand. ∎

REMARK 2.3.7. The result of Theorem 2.3.6 still holds (with a similar proof) if $p\geq 1$, hypothesis (2.3.3) is substituted by the following one

$$f(x,s,z) \geq -a(x,|s|) - b(|s|)\,|z|^p \quad \textit{for every } (x,s,z)\in\Omega\times\mathbf{R}^m\times\mathbf{R}^n$$

where $b(t)$ is increasing in t and $a(x,t)$ is increasing in t and μ-integrable in x, and (i), (ii) are respectively substituted by

(i') *for every* $u\in L^\infty(\Omega;\mathbf{R}^m)$ *and* $v\in L^p(\Omega;\mathbf{R}^n)$ *the functional* $w\to F(u,v+w)$ *is sequentially weakly* l.s.c. on* $L^\infty(\Omega;\mathbf{R}^n)$;

(ii') *for every* $u\in L^\infty(\Omega;\mathbf{R}^m)$ *there exists* $v\in L^p(\Omega;\mathbf{R}^n)$ *such that* $F(u,v)<+\infty$.

2.4. Integral Representation

In this section we consider functionals $F:L^p(\Omega;\mathbf{R}^n)\times\mathfrak{I}\to]-\infty,+\infty]$ and we show that under suitable hypotheses they admit an integral representation of the form

$$F(u,B) = \int_B f(x,u(x))\,d\mu(x) .$$

DEFINITION 2.4.1. *A functional* $F:L^p(\Omega;\mathbf{R}^n)\times\mathfrak{I}\to]-\infty,+\infty]$ *is said*

(a) **local** *on* $\mathfrak{I}$ *if for every* $u,v\in L^p(\Omega;\mathbf{R}^n)$ *and every* $B\in\mathfrak{I}$

$$u=v \;\; \mu\text{-}a.e. \text{ on } B \quad\Rightarrow\quad F(u,B)=F(v,B) ;$$

(b) **additive** *on* $\mathfrak{I}$ *if for every* $u \in L^p(\Omega;\mathbf{R}^n)$ *and every* $B_1,B_2 \in \mathfrak{I}$

$$B_1 \cap B_2 = \emptyset \quad \Rightarrow \quad F(u,B_1 \cup B_2) = F(u,B_1)+F(u,B_2) \ ;$$

(c) **proper** *if there exists* $u_0 \in L^p(\Omega;\mathbf{R}^n)$ *such that*

$$F(u_0,B)<+\infty \qquad \textit{for every } B \in \mathfrak{I} \ .$$

We begin to prove the integral representation result in the case of positive functionals.

THEOREM 2.4.2. *Let* $p \in [1,+\infty[$ *and let* $F: L^p(\Omega;\mathbf{R}^n) \times \mathfrak{I} \to [0,+\infty]$ *be a functional satisfying the following conditions:*

(2.4.1) F *is local on* $\mathfrak{I}$;

(2.4.2) F *is additive on* $\mathfrak{I}$;

(2.4.3) *there exists* $u_0 \in L^p(\Omega;\mathbf{R}^n)$ *such that* $F(u_0,\cdot)$ *is an absolutely continuous measure with respect to* μ;

(2.4.4) *the functional* $F(\cdot,\Omega)$ *is l.s.c. with respect to the strong topology of* $L^p(\Omega;\mathbf{R}^n)$.

Then there exists a positive normal integrand f *such that*

$$(2.4.5) \qquad F(u,B) = \int_B f(x,u(x))\, d\mu(x) \quad \textit{for every } u \in L^p(\Omega;\mathbf{R}^m) \textit{ and } B \in \mathfrak{I}.$$

Moreover, the integrand f *is unique, in the sense that for every integrand* g *satisfying (2.4.5) it is* $g \approx f$.

REMARK 2.4.3. The uniqueness of the representation formula (2.4.5) follows from Corollary 2.1.5. Moreover, in Theorem 2.4.2 it is enough to consider the case $p=1$. In fact, setting $v=u|u|^{p-1}$ and

$$G(v,B) = F(v|v|^{(1-p)/p},B) \qquad \text{for every } v \in L^1(\Omega;\mathbf{R}^m),\ B \in \mathfrak{I}\ ,$$

from the representation of G

$$G(v,B) = \int_B g(x,v)\, d\mu$$

we obtain the representation of F

$$F(u,B) = \int_B f(x,u)\, d\mu$$

with $f(x,s) = g(x,s|s|^{p-1})$. Therefore in the following we consider only the case $p=1$.

For every $k \in \mathbf{N}$ define

$$F_k(u,B) = \inf \left\{ F(v,B) + k\int_B |u-v|\, d\mu \ :\ v \in L^1(\Omega;\mathbf{R}^m) \right\}\ .$$

By using Proposition 1.3.7 we obtain for every $u,v \in L^1(\Omega;\mathbf{R}^m)$ and every $B \in \mathfrak{I}$

$$0 \le F_k(u,B) \le F(u_0,B) + k\int_B |u-u_0|\, d\mu \tag{2.4.6}$$

$$|F_k(u,B) - F_k(v,B)| \le k\int_B |u-v|\, d\mu\ . \tag{2.4.7}$$

PROPOSITION 2.4.4. *For every* $k \in \mathbf{N}$ *there exists a Carathéodory integrand* $f_k(x,s)$ *such that*

$$F_k(u,B) = \int_B f_k(x,u)\, d\mu \qquad \textit{for every } u \in L^1(\Omega;\mathbf{R}^m),\ B \in \mathfrak{I}\ .$$

Proof. Fix $k \in \mathbf{N}$. It is easy to check that for every $s \in \mathbf{R}^m$ the set function $F_k(s,\cdot)$ defined on $\mathfrak{I}$ is additive; moreover, by (2.4.6) it is bounded from above by a measure which is finite and absolutely continuous with respect to μ. Then $F_k(s,\cdot)$ is a measure which is absolutely continuous with respect to μ. Hence, for every $s\in\mathbf{R}^m$ there exists an $\mathfrak{I}$-measurable density $f_{k,s}$ such that

$$(2.4.8)\qquad F_k(s,B) = \int_B f_{k,s}(x)\, d\mu(x) \qquad \text{for every } B\in\mathfrak{I} .$$

From (2.4.6) and (2.4.7) it follows that there exists $N\in\mathfrak{I}$ with $\mu(N)=0$ such that

$$0 \le f_{k,s}(x) \le a(x) + k\,|s-u_0(x)|$$

$$|f_{k,r}(x) - f_{k,s}(x)| \le k\,|r-s|$$

for every $x\in\Omega-N$, and every $r,s\in\mathbf{Q}^m$, where $a\in L^1(\Omega)$ is such that

$$F(u_0,B) = \int_B a(x)\, d\mu(x) \qquad \text{for every } B\in\mathfrak{I} .$$

Fix $x\in\Omega-N$; since the function $s\to f_{k,s}(x)$ is Lipschitz continuous on $\mathbf{Q}^m$,there exists a Lipschitz continuous function on $\mathbf{R}^m$, denoted by $s\to g_{k,s}(x)$, such that $f_{k,s}(x)=g_{k,s}(x)$ for every $s\in\mathbf{Q}^m$. Define now for every $x\in\Omega$ and $s\in\mathbf{R}^m$

$$(2.4.9)\qquad f_k(x,s) = \begin{cases} g_{k,s}(x) & \text{if } x\in\Omega-N \\ 0 & \text{if } x\in N . \end{cases}$$

It is easy to check that the function f_k is a Carathéodory integrand, and that

$$0 \le f_k(x,s) \le a(x) + k\,|s-u_0(x)| \qquad \text{for every } x\in\Omega \text{ and } s\in\mathbf{R}^m.$$

For every $u\in L^1(\Omega;\mathbf{R}^m)$ and $B\in\mathfrak{I}$ set

$$G(u,B) = \int_B f_k(x,u)\, d\mu .$$

From (2.4.1), (2.4.2), (2.4.8), (2.4.9) it follows that

$$G(u,B) = F_k(u,B)$$

for every $B\in\mathfrak{I}$ and every simple function u which takes its values in $\mathbf{Q}^m$. The functionals $G(\cdot,B)$ and $F_k(\cdot,B)$ are $L^1(\Omega;\mathbf{R}^m)$-continuous for every $B\in\mathfrak{I}$ and coincide on a dense subset of $L^1(\Omega;\mathbf{R}^m)$. Thus, they coincide on the whole space $L^1(\Omega;\mathbf{R}^m)$, so that

$$F_k(u,B) = \int_B f_k(x,u)\, d\mu \qquad \text{for every } u \in L^1(\Omega;\mathbf{R}^m) \text{ and } B \in \mathfrak{I}\ ,$$

and the proof is achieved. ∎

Proof of Theorem 2.4.2. By Proposition 2.4.4 for every $k \in \mathbf{N}$ there exists a Carathéodory integrand $f_k(x,s)$ such that

$$(2.4.10) \qquad F_k(u,B) = \int_B f_k(x,u)\, d\mu \quad \text{for every } u \in L^1(\Omega;\mathbf{R}^m),\ B \in \mathfrak{I}.$$

Since $0 \le F_k(u,B) \le F_{k'}(u,B)$ for $k \le k'$, by Proposition 2.1.3 there exists $N \in \mathfrak{I}$ with $\mu(N)=0$ such that

$$0 \le f_k(x,s) \le f_{k'}(x,s)$$

for every $x \in \Omega - N$, $s \in \mathbf{R}^m$, $k,k' \in \mathbf{N}$ with $k \le k'$. Set now

$$(2.4.11) \qquad f(x,s) = \begin{cases} \sup\limits_{k \in \mathbf{N}} f_k(x,s) & \text{if } x \in \Omega - N \\ 0 & \text{if } x \in N\,. \end{cases}$$

The function f is a normal integrand, because it is the supremum of an increasing sequence of Carathéodory integrands. Moreover, since $F(\cdot,\Omega)$ is $L^1(\Omega;\mathbf{R}^m)$-l.s.c., by (2.4.1), (2.4.2), (2.4.3) it follows that $F(\cdot,B)$ is $L^1(\Omega;\mathbf{R}^m)$-l.s.c. for every $B \in \mathfrak{I}$. This implies, by Proposition 1.3.7(ii)

$$F(u,B) = \sup_{k \in \mathbf{N}} F_k(u,B) \qquad \text{for every } u \in L^1(\Omega;\mathbf{R}^m),\ B \in \mathfrak{I}\ .$$

By (2.4.10), (2.4.11), and by the monotone convergence theorem, we obtain

$$F(u,B) = \int_B f(x,u)\, d\mu \qquad \text{for every } u \in L^1(\Omega;\mathbf{R}^m),\ B \in \mathfrak{I}\ ,$$

so that the theorem is proved. ■

The case of functionals $F:L^p(\Omega;\mathbf{R}^m)\times\mathfrak{I}\to]-\infty,+\infty]$ can be reduced to the case of positive functionals studied in Theorem 2.4.2 by using the following lemma.

LEMMA 2.4.5. *Let* $F:L^p(\Omega;\mathbf{R}^m)\times\mathfrak{I}\to]-\infty,+\infty]$ *be a functional such that (2.4.1), (2.4.2), (2.4.3), (2.4.4) hold. Then, for every* $u\in L^p(\Omega;\mathbf{R}^m)$ *the set function* $F(u,\cdot)$ *is a signed measure on* $\mathfrak{I}$ *(not necessarily σ-finite), and there exist* $a\in L^1(\Omega)$ *and* $b\geq 0$ *such that*

$$F(u,B) \geq -\int_B \left[a(x) + b|u|^p\right] d\mu \qquad \textit{for every } u\in L^p(\Omega;\mathbf{R}^m),\ B\in\mathfrak{I}\ .$$

Proof. By considering the functional

$$G(u,B) = F(u+u_0,B) - F(u_0,B)$$

(where u_0 is the function given by (2.4.3)), we may restrict ourselves to the case when

(2.4.12) $\qquad F(0,B) = 0 \qquad$ for every $B\in\mathfrak{I}$.

Therefore, since $F(\cdot,\Omega)$ is $L^p(\Omega;\mathbf{R}^m)$-l.s.c., for every $\varepsilon>0$ there exists $\delta(\varepsilon)>0$ such that

(2.4.13) $\qquad F(u,\Omega) > -\varepsilon \qquad$ whenever $u\in L^p(\Omega;\mathbf{R}^m)$ with $\int_\Omega |u|^p\, d\mu < \delta(\varepsilon)$.

Let us prove that

(2.4.14) $$F(u,B) + \frac{\varepsilon}{\delta(\varepsilon)}\int_B |u|^p\, d\mu \geq -\varepsilon$$

for every $\varepsilon>0$, $u\in L^p(\Omega;\mathbf{R}^m)$, $B\in\mathfrak{I}$. The proof is a modification of an argument of

Vainberg (see [282], Theorem 19.1). Fix $\varepsilon>0$, $u\in L^p(\Omega;\mathbf{R}^m)$, $B\in\mathfrak{I}$; for suitable $k\in\mathbf{N}$ and $\eta\in[0,1[$ we have

$$\int_B |u|^p\,d\mu = (k+\eta)\delta(\varepsilon)\ . \tag{2.4.15}$$

Since μ is non-atomic, there exist $k+1$ sets $B_1,\dots,B_{k+1}$ belonging to $\mathfrak{I}$, pairwise disjoint and such that

$$B = \bigcup_{i=1}^{k+1} B_i \qquad \text{and} \qquad \int_{B_i} |u|^p\,d\mu < \delta(\varepsilon) \quad \text{for } i=1,\dots,k+1\ .$$

By (2.4.13) we have

$$F(u,B_i) = F(u\cdot 1_{B_i},\Omega) > -\varepsilon \qquad\qquad \text{for } i=1,\dots,k+1\ ,$$

which implies

$$F(u,B) = \sum_{i=1}^{k+1} F(u,B_i) \geq -(k+1)\varepsilon\ .$$

Therefore, by (2.4.15)

$$F(u,B) + \frac{\varepsilon}{\delta(\varepsilon)}\int_B |u|^p\,d\mu \geq -(k+1)\varepsilon + (k+\eta)\varepsilon \geq -\varepsilon\ ,$$

which proves (2.4.14).

We prove now that for every $u\in L^p(\Omega;\mathbf{R}^m)$ the set function $F(u,\cdot)$ is a signed measure on $(\Omega,\mathfrak{I})$. Fix $u\in L^p(\Omega;\mathbf{R}^m)$; since $F(u,\cdot)$ is finitely additive on $\mathfrak{I}$, it is enough to prove that

$$F(u,B) = \lim_{h\to\infty} F(u,B_h) \qquad \text{whenever } B_h\uparrow B\ . \tag{2.4.16}$$

Fix $B_h\uparrow B$; by the lower semicontinuity of $F(\cdot,\Omega)$ we have

$$\begin{aligned} F(u,B) &= F(u\cdot 1_B,\Omega) \leq \\ &\leq \liminf_{h\to\infty} F(u\cdot 1_{B_h},\Omega) = \liminf_{h\to\infty} F(u,B_h)\ . \end{aligned} \tag{2.4.17}$$

On the other hand, from (2.4.14) we obtain for every $\varepsilon>0$

$$F(u,B)+\frac{\varepsilon}{\delta(\varepsilon)}\int_B |u|^p\,d\mu =$$

$$= F(u,B_h)+\frac{\varepsilon}{\delta(\varepsilon)}\int_{B_h} |u|^p\,d\mu + F(u,B-B_h)+\frac{\varepsilon}{\delta(\varepsilon)}\int_{B-B_h} |u|^p\,d\mu \geq$$

$$\geq F(u,B_h)+\frac{\varepsilon}{\delta(\varepsilon)}\int_{B_h} |u|^p\,d\mu - \varepsilon\ .$$

Therefore,

$$F(u,B)+\frac{\varepsilon}{\delta(\varepsilon)}\int_B |u|^p\,d\mu \geq \limsup_{h\to\infty}\Big[F(u,B_h)+\frac{\varepsilon}{\delta(\varepsilon)}\int_{B_h} |u|^p\,d\mu\Big] - \varepsilon\ ,$$

so that

$$F(u,B) \geq \limsup_{h\to\infty} F(u,B_h) - \varepsilon\ ,$$

and, since $\varepsilon>0$ was arbitrary, by (2.4.17) we obtain (2.4.16).

For every $\varepsilon>0$, $u\in L^p(\Omega;\mathbf{R}^m)$, $B\in\mathfrak{I}$ set

$$\Phi_\varepsilon(u,B) = -F(u,B)-\frac{\varepsilon}{\delta(\varepsilon)}\int_B |u|^p\,d\mu$$

$$\alpha_\varepsilon(B) = \sup\left\{\Phi_\varepsilon(u,B')\ :\ B'\in\mathfrak{I},\ B'\subset B,\ u\in L^p(\Omega;\mathbf{R}^m)\right\}.$$

Let us prove that α_ε is a positive measure on $\mathfrak{I}$. It is easy to check that α_ε is increasing and finitely additive on $\mathfrak{I}$; moreover, from (2.4.14) it follows that

$$0\leq\alpha_\varepsilon(B)\leq\varepsilon \qquad \text{for every } B\in\mathfrak{I}\ .$$

It remains to prove that $\alpha_\varepsilon(B_h)\uparrow\alpha_\varepsilon(B)$ whenever $B_h\uparrow B$. By monotonicity, we have

$$\alpha_\varepsilon(B) \geq \sup_{h\in\mathbf{N}} \alpha_\varepsilon(B_h)\ ;$$

conversely, if $t<\alpha_\varepsilon(B)$, by the definition of α_ε there exist $u\in L^p(\Omega;\mathbf{R}^m)$ and

$B' \in \mathfrak{I}$ such that $B' \subset B$ and $t < \Phi_\varepsilon(u,B')$. Since $F(u,\cdot)$ is a measure, we have

$$\Phi_\varepsilon(u,B') = \lim_{h\to\infty} \Phi_\varepsilon(u,B' \cap B_h) ,$$

so that

$$t < \lim_{h\to\infty} \Phi_\varepsilon(u,B' \cap B_h) \le \sup_{h\in \mathbf{N}} \alpha_\varepsilon(B_h) .$$

Since $t < \alpha_\varepsilon(B)$ was arbitrary, this yields

$$\alpha_\varepsilon(B) \le \sup_{h\in \mathbf{N}} \alpha_\varepsilon(B_h) .$$

Therefore α_ε is a bounded positive measure on $\mathfrak{I}$.

By (2.4.12) it follows that $F(u,B)=0$ whenever $\mu(B)=0$, so that

$$\alpha_\varepsilon(B) = 0 \qquad \text{for every } B \in \mathfrak{I} \text{ with } \mu(B)=0 .$$

Then, by the Radon-Nikodym theorem, there exist $a_\varepsilon \in L^1(\Omega)$ such that

$$\alpha_\varepsilon(B) = \int_B a_\varepsilon(x)\, d\mu \qquad \text{for every } B \in \mathfrak{I} .$$

By the definition of α_ε we have, for every $u \in L^p(\Omega;\mathbf{R}^m)$ and $B \in \mathfrak{I}$

$$-F(u,B) - \frac{\varepsilon}{\delta(\varepsilon)} \int_B |u|^p\, d\mu \le \alpha_\varepsilon(B) = \int_B a_\varepsilon(x)\, d\mu ,$$

which yields

$$F(u,B) \ge -\int_B \Big[a_\varepsilon(x) + \frac{\varepsilon}{\delta(\varepsilon)} |u|^p\Big]\, d\mu$$

for every $\varepsilon > 0$, $u \in L^p(\Omega;\mathbf{R}^m)$, $B \in \mathfrak{I}$. ∎

We are now in a position to prove the integral representation theorem in its general form.

THEOREM 2.4.6. *Let* $F: L^p(\Omega;\mathbf{R}^m) \times \mathfrak{I} \to]-\infty,+\infty]$ *be a functional such that*

F *is local on* $\mathfrak{I}$;

F *is additive on* $\mathfrak{I}$;

F *is proper*;

$F(\cdot,\Omega)$ *is* $L^p(\Omega;\mathbf{R}^m)$*-l.s.c.*.

Then, there exists a normal integrand $f(x,s)$ *such that*

(i) $f(x,s) \geq -a(x) - b|s|^p$ *for* μ*-a.e.* $x \in \Omega$ *and every* $s \in \mathbf{R}^m$;

(ii) $F(u,B) - F(u_0,B) = \int_B f(x,u)\, d\mu$ *for every* $u \in L^p(\Omega;\mathbf{R}^m)$, $B \in \mathfrak{I}$

where $a(x)$ *is a suitable function in* $L^1(\Omega)$, $b \geq 0$, *and* u_0 *is given by Definition 2.4.1(c). Moreover, the integrand* f *is unique, in the sense that for every integrand* g *satisfying (ii) we have* $g \approx f$.

Proof. The functional

$$G(u,B) = F(u,B) - F(u_0,B)$$

satisfies all conditions of Lemma 2.4.5, so that

$$G(u,B) \geq -\int_B \left[a(x) + b|u|^p\right] d\mu \qquad \text{for every } u \in L^p(\Omega;\mathbf{R}^m),\ B \in \mathfrak{I}$$

for suitable $a \in L^1(\Omega)$ and $b \geq 0$. Consider now the functional

$$H(u,B) = G(u,B) + \int_B \left[a(x) + b|u|^p\right] d\mu \ ;$$

by Theorem 2.4.2 we have

$$H(u,B) = \int_B h(x,u)\, d\mu \qquad \text{for every } u \in L^p(\Omega;\mathbf{R}^m),\ B \in \mathfrak{I}$$

for a suitable non-negative normal integrand $h(x,s)$. Therefore, setting

$$f(x,s) = h(x,s) - a(x) - b|s|^p \ ,$$

we have that f is a normal integrand which satisfies the required conditions (i) and (ii). The uniqueness of f follows from Corollary 2.1.5. ■

COROLLARY 2.4.7. *Let* $F:L^p(\Omega;\mathbf{R}^m)\times\mathfrak{I}\to\mathbf{R}$ *be a functional local and additive on* $\mathfrak{I}$, *and such that*

$$F(\cdot,\Omega) \text{ is } L^p(\Omega;\mathbf{R}^m)\text{-continuous.}$$

Then, the integrand f *given by Theorem 2.4.6 is a Carathéodory integrand. Moreover, for suitable* $a\in L^1(\Omega)$ *and* $b\geq 0$ *we have*

$$|f(x,s)| \leq a(x) + b|s|^p \qquad \text{for } \mu\text{-a.e. } x\in\Omega \text{ and for all } s\in\mathbf{R}^m.$$

Proof. It is enough to apply Theorem 2.4.6 to F and to –F taking into account the uniqueness of the integral representation guaranteed by Corollary 2.1.5. ■

COROLLARY 2.4.8. *Assume that* $F:L^p(\Omega;\mathbf{R}^m)\times\mathfrak{I}\to]-\infty,+\infty]$ *is local, additive, proper, and that*

$$F(\cdot,\Omega) \text{ is seq. } wL^p(\Omega;\mathbf{R}^m)\text{-l.s.c..}$$

Then, the integrand f *given by Theorem 2.4.6 is a convex integrand.*

Proof. It is enough to apply Theorem 2.4.6 and Remark 2.3.7. ■

2.5. A Counterexample

Coming back to the statement of Theorem 2.4.6, we remark that the locality and additivity hypotheses are necessary in order to get an integral representation formula like

$$F(u,B) = \int_B f(x,u)\, d\mu \tag{2.5.1}$$

where $f(x,s)$ is a suitable integrand, and the hypothesis that F is proper is not very restrictive. As regards to the lower semicontinuity of $F(\cdot,\Omega)$, one may think it is not necessary to obtain the integral representation (2.5.1). In this section we show by a counterexample that if the lower semicontinuity hypothesis is dropped, then the integral representation formula (2.5.1) may fail.

In all this section we shall take $\Omega=]0,1[$, $\mathfrak{I}$ the σ-field of all Lebesgue measurable subsets of Ω, μ the Lebesgue measure on Ω, $n=1$, and p any number in $[1,+\infty[$. For every $u\in L^p(\Omega)$ we denote by $Tu:\Omega\to\mathbf{R}$ the function defined by

$$(Tu)(x) = \begin{cases} 1 & \text{if } \mu(\{y\in\Omega : u(y)=u(x)\}) = 0 \\ 0 & \text{otherwise} . \end{cases}$$

Finally, we set for every $u\in L^p(\Omega)$ and $B\in\mathfrak{I}$

$$F(u,B) = \int_B Tu\, d\mu .$$

THEOREM 2.5.1. *The following properties for* F *hold:*

(i) F *is local on* $\mathfrak{I}$ *in the sense of Definition 2.4.1(a);*

(ii) *for every* $u\in L^p(\Omega)$ *the set function* $F(u,\cdot)$ *is a positive measure on* $\mathfrak{I}$;

(iii) *for every* $B\in\mathfrak{I}$ *we have* $F(0,B)=0$;

(iv) *for every integrand* $f(x,s)$ *there exist* $u\in L^p(\Omega)$ *and* $B\in\mathfrak{I}$ *such that*

$$F(u,B) \neq \int_B f(x,u)\, d\mu .$$

Proof. Properties (ii) and (iii) follow immediately from the definition of F. Let us prove the locality condition (i). For every $u \in L^p(\Omega)$ and $B \in \mathfrak{I}$ define

$$N(u,B) = \{t \in \mathbf{R} : \mu(\{y \in B : u(y)=t\}) \neq 0\} .$$

The set $N(u,B)$ is at most countable, and we have

$$u=v \text{ a.e. on } B \quad \Rightarrow \quad N(u,B)=N(v,B) .$$

Moreover, the function Tu can be written in the form

$$(Tu)(x) = \begin{cases} 1 & \text{if } u(x) \notin N(u,\Omega) \\ 0 & \text{otherwise} . \end{cases}$$

For every $u \in L^p(\Omega)$ and $B \in \mathfrak{I}$ we define the function $T^B u : \Omega \to \mathbf{R}$ by

$$(T^B u)(x) = \begin{cases} 1 & \text{if } u(x) \notin N(u,B) \\ 0 & \text{otherwise} . \end{cases}$$

In order to prove (i) it is enough to show that

(2.5.2) for every $u \in L^p(\Omega)$ and $B \in \mathfrak{I}$ we have $Tu=T^B u$ a.e. on B.

Fix $u \in L^p(\Omega)$ and $B \in \mathfrak{I}$; since $N(u,B) \subset N(u,\Omega)$ we have

$$\{x \in B : (Tu)(x) \neq (T^B u)(x)\} = \{x \in B : u(x) \in N(u,\Omega)-N(u,B)\} =$$

$$= \bigcup_{t \in N(u,\Omega)-N(u,B)} \{x \in B : u(x)=t\} .$$

Since $N(u,\Omega)$ is at most countable and $\mu(\{x \in B : u(x)=t\})=0$ for every real number $t \in N(u,\Omega)-N(u,B)$, we have

$$\mu(\{x \in B : (Tu)(x) \neq (T^B u)(x)\}) = \sum_{t \in N(u,\Omega)-N(u,B)} \mu(\{x \in B : u(x)=t\}) = 0 ,$$

and (2.5.2) is proved.

Let us prove condition (iv). We argue by contradiction. Suppose there exists an integrand $f(x,s)$ such that

$$F(u,B) = \int_B f(x,u)\,dx \qquad \text{for every } u \in L^p(\Omega) \text{ and } B \in \mathfrak{I}\,.$$

Consider the family of functions $(u_c)_{c\in\mathbf{R}}$ defined by

$$u_c(x) = x + c\,.$$

For every $c \in \mathbf{R}$ and $B \in \mathfrak{I}$ we have

$$F(u_c,B) = \mu(B)\ ,$$

so that, for every $c \in \mathbf{R}$

$$f(x,x+c) = 1 \qquad \text{a.e. on } \Omega.$$

By Fubini's theorem this yields

$$f(x,s) = 1 \qquad \text{for a.e. } (x,s) \in \Omega\times\mathbf{R}.$$

Hence, by Fubini's theorem again, there exists $s \in \mathbf{R}$ such that

$$f(x,s) = 1 \qquad \text{for a.e. } x \in \Omega\,,$$

and so

$$F(s,B) = \mu(B) \qquad \text{for every } B \in \mathfrak{I},$$

while, by the definition of F, it is

$$F(s,B) = 0 \qquad \text{for every } B \in \mathfrak{I}.$$

This contradiction proves (iv), and the proof of Theorem 2.5.1 is achieved. ∎

REMARK 2.5.2. The counterexample of Theorem 2.5.1 shows that if we drop the lower semicontinuity assumption (2.4.4), then the functional F may be not representable by the integral of a suitable integrand $f(x,s)$. However, it is possible to show that, under the Continuum Hypothesis, every local, additive, and proper functional F admit a representation of the form

$$F(u,B) = F(u_0,B) + \int_B f(x,u)\,d\mu \qquad \text{for every } u \in L^p(\Omega;\mathbf{R}^m) \text{ and } B \in \mathfrak{I}$$

with f(x,s) possibly non $\Im\otimes\mathbb{B}_m$-measurable, but such that for every $u\in L^p(\Omega;\mathbf{R}^m)$ the function f(x,u(x)) is μ-measurable. For further details on results of this kind, we refer to Appell [19].

2.6. Relaxation in L^p Spaces

In this section we deal with the relaxation problem for functionals of the form

$$F(u,v) = \int_\Omega^* f(x,u,v)\,d\mu \qquad u\in L^p(\Omega;\mathbf{R}^m),\ v\in L^q(\Omega;\mathbf{R}^n)$$

where $p,q\in[1,+\infty[$, $f:\Omega\times\mathbf{R}^m\times\mathbf{R}^n\to[0,+\infty]$ is a function, and $\int^*$ denotes the upper integral defined in Section 2.1. For simplicity we consider only the case $f\geq 0$. The problem we are interested in is the characterization into an integral form of the relaxed functional

$$\Gamma F(u,v) = \Gamma_{seq}\big((L^p\times wL^q)^-\big)\,F(u,v)\ .$$

More precisely, we want to show that

$$\Gamma F(u,v) = \int_\Omega g(x,u,v)\,d\mu$$

for a suitable integrand g(x,s,z), and characterize g in terms of f. To do this, it is convenient to localize the functionals F and ΓF by setting for all $u\in L^p(\Omega;\mathbf{R}^m)$, $v\in L^q(\Omega;\mathbf{R}^n)$, $B\in\Im$

$$(2.6.1)\qquad F(u,v,B) = \int_B^* f(x,u,v)\,d\mu$$

(2.6.2) $$\Gamma F(u,v,B) = \Gamma_{seq}\big((L^p \times wL^q)^-\big)\, F(u,v,B)\ .$$

PROPOSITION 2.6.1. *The localized functional* ΓF *satisfies the following conditions:*

(i) *for every* $B \in \mathfrak{I}$ *the function* $\Gamma F(\cdot,\cdot,B)$ *is seq.* $L^p \times wL^q$- *l.s.c. (hence* $L^p \times L^q$-*l.s.c.);*

(ii) ΓF *is local on* $\mathfrak{I}$ *in the sense of Definition 2.4.1(a);*

(iii) *for every* $u \in L^p(\Omega;\mathbf{R}^m)$ *and* $v \in L^q(\Omega;\mathbf{R}^n)$ *the set function* $\Gamma F(u,v,\cdot)$ *is additive on* $\mathfrak{I}$.

Proof. Assertion (i) follows immediately from Proposition 1.3.1(i).

We prove assertion (ii) by using Proposition 1.3.2. If Λ is the set of all countable ordinals and F_λ are defined as in Proposition 1.3.2, it is enough to show that

(2.6.3) F_λ is local on $\mathfrak{I}$ for every $\lambda \in \Lambda$.

We prove (2.6.3) by transfinite induction. When $\lambda=0$, the functional $F_0=F$ is clearly local on $\mathfrak{I}$. Assume now that for a given $\lambda \in \Lambda$ the functional F_λ is local on $\mathfrak{I}$, and let $U_1=(u_1,v_1)$ and $U_2=(u_2,v_2)$ be two elements of $L^p(\Omega;\mathbf{R}^m) \times L^q(\Omega;\mathbf{R}^n)$ such that $U_1=U_2$ μ-a.e. on a certain $B \in \mathfrak{I}$. We shall prove that

(2.6.4) $F_{\lambda+1}(U_1,B) \le F_{\lambda+1}(U_2,B)$

the proof of the opposite inequality being similar. Let $\{U_{2,h}\}$ be a sequence converging to U_2 in the topology $L^p \times wL^q$ and let for every $h \in \mathbf{N}$

$$U_{1,h} = 1_B\, U_{2,h} + 1_{\Omega - B}\, U_1\ .$$

The sequence $\{U_{1,h}\}$ converges to U_1 in the topology $L^p \times wL^q$ and, by the locality of F_λ we have

$$F_{\lambda+1}(U_1,B) \le \liminf_{h\to+\infty} F_\lambda(U_{1,h},B) = \liminf_{h\to+\infty} F_\lambda(U_{2,h},B) .$$

Since the sequence $\{U_{2,h}\}$ was arbitrary, (2.6.4) is proved. Finally, if λ is a limit ordinal and F_α is local on $\mathfrak{I}$ for every $\alpha<\lambda$, the locality of F_λ follows easily.

We prove now assertion (iii). Again by Proposition 1.3.2 it is enough to show that

(2.6.5) $\quad F_\lambda$ is additive on $\mathfrak{I}$ for every $\lambda\in\Lambda$.

As before, we prove (2.6.5) by transfinite induction. When $\lambda=0$, the functional $F_0=F$ is clearly additive on $\mathfrak{I}$. Let us show that

(2.6.6) $\quad F_\lambda$ additive $\Rightarrow$ $F_{\lambda+1}$ additive.

Fix $\lambda\in\Lambda$, $U=(u,v)\in L^p(\Omega;\mathbf{R}^m)\times L^q(\Omega;\mathbf{R}^n)$, and $B_1,B_2\in\mathfrak{I}$ with $B_1\cap B_2=\emptyset$. If $\{U_h\}$ is a sequence converging to U in the topology $L^p\times wL^q$ we have

$$F_{\lambda+1}(U,B_1) + F_{\lambda+1}(U,B_2) \le \liminf_{h\to+\infty} F_\lambda(U_h,B_1) + \liminf_{h\to+\infty} F_\lambda(U_h,B_2) \le$$

$$\le \lim_{h\to+\infty} F_\lambda(U_h,B_1\cup B_2) ,$$

so that, since $\{U_h\}$ was arbitrary,

$$F_{\lambda+1}(U,B_1) + F_{\lambda+1}(U,B_2) \le F_{\lambda+1}(U,B_1\cup B_2) .$$

In order to prove the opposite inequality, assume $F_{\lambda+1}(U,B_1)$ and $F_{\lambda+1}(U,B_2)$ are finite, and for every $\varepsilon>0$ let $\{U_{1,h}\}$ and $\{U_{2,h}\}$ be two sequences converging to U in the topology $L^p\times wL^q$ such that

$$\lim_{h\to+\infty} F_\lambda(U_{1,h},B_1) \le F_{\lambda+1}(U,B_1) + \varepsilon$$

$$\lim_{h\to+\infty} F_\lambda(U_{2,h},B_2) \le F_{\lambda+1}(U,B_2) + \varepsilon .$$

Set for every $h\in\mathbf{N}$

$$U_h = 1_{B_1} U_{1,h} + 1_{\Omega-B_1} U_{2,h} .$$

The sequence $\{U_h\}$ converges to U in the topology $L^p\times wL^q$ and, by using the locality and the additivity of F_λ we have

$$F_{\lambda+1}(U,B_1\cup B_2) \le \liminf_{h\to+\infty} F_\lambda(U_h,B_1\cup B_2) =$$

$$= \lim_{h\to+\infty} \left[F_\lambda(U_{1,h},B_1) + F_\lambda(U_{2,h},B_2)\right] \le$$

$$\le F_{\lambda+1}(U,B_1) + F_{\lambda+1}(U,B_2) + 2\varepsilon \ .$$

Since $\varepsilon>0$ was arbitrary, we have

$$F_{\lambda+1}(U,B_1\cup B_2) \le F_{\lambda+1}(U,B_1) + F_{\lambda+1}(U,B_2) \ ,$$

and (2.6.6) is proved. Finally, if λ is a limit ordinal, the additivity of F_λ follows easily from the additivity of F_α for every $\alpha<\lambda$. ■

THEOREM 2.6.2. *Let* F *and* ΓF *be the functionals defined in (2.6.1) and (2.6.2). Assume that for suitable* $u_0\in L^p(\Omega;\mathbf{R}^m)$ *and* $v_0\in L^q(\Omega;\mathbf{R}^n)$ *we have*

$$F(u_0,v_0,\Omega) < +\infty \ .$$

Then there exists a normal-convex integrand $g(x,s,z)$ *such that*

$$\Gamma F(u,v,B) = \int_B g(x,u,v)\, d\mu$$

for every $u\in L^p(\Omega;\mathbf{R}^m)$, $v\in L^q(\Omega;\mathbf{R}^n)$, *and* $B\in\mathfrak{I}$.

Proof. By Proposition 2.1.7 there exists $\phi\in L^1(\Omega)$ such that

$$\int_B^* f(x,u_0,v_0)\, d\mu = \int_B \phi\, d\mu \qquad \text{for every } B\in\mathfrak{I} \ .$$

Since

$$\Gamma F(u_0,v_0,B) \le \int_B \phi\, d\mu \qquad \text{for every } B\in\mathfrak{I} \ ,$$

the set function $\Gamma F(u_0,v_0,\cdot)$ is actually a measure which is absolutely continuous with respect to μ. Therefore, taking into account Proposition 2.6.1, by the integral representation Theorem 2.4.2 we get

$$\Gamma F(u,v,B) = \int_B g(x,u,v)\, d\mu$$

for every $u \in L^p(\Omega;\mathbf{R}^m)$, $v \in L^q(\Omega;\mathbf{R}^n)$, and $B \in \mathfrak{I}$, where $g(x,s,z)$ is a suitable normal integrand. The fact that $g(x,s,z)$ is convex in z follows from Theorem 2.3.6 and Remark 2.3.7. ■

In order to characterize the integrand $g(x,s,z)$, for any function $f:\mathbf{R}^m \times \mathbf{R}^n \to \overline{\mathbf{R}}$ we set

(2.6.7) $\quad \underline{co}\, f(s,z) = \sup\Big\{ g(s,z) \ :\ g \leq f,\ g(s,z) \text{ is l.s.c. in } (s,z) \text{ and convex in } z \Big\}.$

Since the supremum of a family of convex (respectively l.s.c.) functions is still convex (respectively l.s.c.), we have that the supremum in (2.6.7) is actually attained. Note that when f does not depend on s it is

$$\underline{co}\, f(z) = f^{**}(z)$$

and when f does not depend on z it is

$$\underline{co}\, f(s) = \Gamma f(s)$$

where $f^{**}(z)$ and $\Gamma f(s)$ respectively denote the convex l.s.c. envelope of f in $\mathbf{R}^n$ and the l.s.c. envelope of f in $\mathbf{R}^m$.

PROPOSITION 2.6.3. *Let* $f:\Omega \times \mathbf{R}^m \times \mathbf{R}^n \to [0,+\infty]$ *be a positive integrand. Then the function* $\underline{co}\, f(x,s,z)$ *defined by*

$$\underline{co}\, f(x,\cdot,\cdot) = \underline{co}\, [f(x,\cdot,\cdot)]$$

is a normal-convex integrand.

Proof. By Lemma 2.1.6 and Proposition 2.1.7 there exists a normal integrand $g(x,s,z)$ such that

(2.6.8) $\underline{co}\, f(x,s,z) \leq g(x,s,z)$ for every $x\in\Omega$, $s\in\mathbf{R}^m$, $z\in\mathbf{R}^n$;

(2.6.9) $\int^{*}_{\Omega} \underline{co}\, f(x,u,v)\, d\mu = \int_{\Omega} g(x,u,v)\, d\mu$ for every $u\in L^1(\Omega;\mathbf{R}^m)$, $v\in L^1(\Omega;\mathbf{R}^n)$.

Fix $u\in L^1(\Omega;\mathbf{R}^m)$; since $\underline{co}\, f(x,s,z)$ is convex in z, by (2.6.9) the functional

(2.6.10) $$v \rightarrow \int_{\Omega} g(x,u(x),v)\, d\mu$$

is convex. Moreover, since g is a normal integrand, the functional (2.6.10) is $L^1(\Omega;\mathbf{R}^n)$-l.s.c., hence $wL^1(\Omega;\mathbf{R}^n)$-l.s.c. by Proposition 1.1.6(i). Therefore, by Theorem 2.3.6 and Remark 2.3.7 the function

$$g_u(x,z) = g(x,u(x),z)$$

turns out to be a convex integrand, so that by Lemma 2.1.10 the function $g(x,s,z)$ is a normal-convex integrand. By Lemma 2.1.6 it is $g \lesssim f$, hence by the definition of $\underline{co}\, f$ we have

$$g \lesssim \underline{co}\, f$$

which, together with (2.6.8) achieves the proof. ■

When the function f in (2.6.1) is an integrand, we may characterize the function g given by Theorem 2.6.2. More precisely the following result holds.

THEOREM 2.6.4. *Let* $f:\Omega\times\mathbf{R}^m\times\mathbf{R}^n\rightarrow[0,+\infty]$ *be a positive integrand, and let* F *and* ΓF *be the functionals defined in (2.6.1) and (2.6.2). Assume that for suitable* $u_0\in L^p(\Omega;\mathbf{R}^m)$ *and* $v_0\in L^q(\Omega;\mathbf{R}^n)$ *we have* $F(u_0,v_0,\Omega)<+\infty$. *Then, for every* $u\in L^p(\Omega;\mathbf{R}^m)$, $v\in L^q(\Omega;\mathbf{R}^n)$, $B\in\mathfrak{I}$ *we have*

$$\Gamma F(u,v,B) = \int_B \underline{co}\, f(x,u,v)\, d\mu \ .$$

Proof. By Theorem 2.6.2 the functional ΓF can be represented in the form

$$\Gamma F(u,v,B) = \int_B g(x,u,v)\, d\mu$$

for a suitable normal-convex integrand g. Since $\Gamma F \leq F$, by Proposition 2.1.3 we have $g \lesssim f$, hence $g \lesssim \underline{co}\, f$. On the other hand, by Proposition 2.6.3 $\underline{co}\, f$ is a normal-convex integrand, hence by the lower semicontinuity Theorem 2.3.1 the functional

$$\int_B \underline{co}\, f(x,u,v)\, d\mu$$

is sequentially l.s.c. with respect to $L^p(\Omega;\mathbf{R}^m) \times wL^q(\Omega;\mathbf{R}^n)$. Then we have

$$\int_B \underline{co}\, f(x,u,v)\, d\mu \ \leq \ \Gamma F(u,v,B) \ = \ \int_B g(x,u,v)\, d\mu \ ,$$

so that Proposition 2.1.3 yields $\underline{co}\, f \lesssim g$. Therefore, we have proved that $\underline{co}\, f \approx g$, and so

$$\Gamma F(u,v,B) = \int_B \underline{co}\, f(x,u,v)\, d\mu \ . \blacksquare$$

REMARK 2.6.5. Under the assumptions of Theorem 2.6.4, if $f=f(x,z)$ we have

$$\Gamma_{seq}(wL^q(\Omega;\mathbf{R}^n)^-)\, F(v,B) \ = \ \int_B f^{**}(x,v)\, d\mu \ ,$$

and if $f=f(x,s)$

$$\Gamma(L^p(\Omega;\mathbf{R}^m)^-)\, F(u,B) \ = \ \int_B \Gamma f(x,u)\, d\mu \ ,$$

where $f^{**}(x,z)$ and $\Gamma f(x,s)$ respectively denote the convex l.s.c. envelope of f

with respect to z and the l.s.c. envelope of f with respect to s.

2.7. Further Remarks

Lower semicontinuity results similar to the ones of Section 2.3 have been considered by many authors in different frameworks; let us recall an unpublished paper by De Giorgi (see [126]), and, among the others, Berkowitz [47], Cesari [91], Olech [240].

As we already mentioned, the lower semicontinuity results of Section 2.3 have been proved in that form by Ioffe in [183]; recent generalizations have been obtained by Bottaro & Oppezzi (see [55]) when the integrand f is defined on $\Omega \times X \times Y$ where X and Y are Banach spaces, and by Ambrosio (see [9]) and Balder (see [28]) when no measurability hypotheses are required on f (see also Theorem 2.3.3). In this last case the functional F is of the form

$$F(u,v) = \int_{\Omega}^{*} f(x,u,v)\, d\mu$$

where $\int^*$ denotes the upper integral defined in Section 2.1.

Concerning the inf-compactness property (ii) of Theorem 2.3.1, in Ioffe's paper [183] it is shown it is necessary in order that the functional

$$F(u,v) = \int_{\Omega} f(x,u,v)\, d\mu$$

be well-defined, sequentially $L^p(\Omega;\mathbf{R}^m) \times wL^q(\Omega;\mathbf{R}^n)$-l.s.c., and greater than $-\infty$ on $L^p(\Omega;\mathbf{R}^m) \times L^q(\Omega;\mathbf{R}^n)$.

The integral representation results of Section 2.4 have been proved by Buttazzo & Dal Maso in [73] and, with a different proof by Hiai in [180]. Further generalizations to Orlicz spaces have been obtained by Fougères & Truffert (see [158]); for a wide review on the subject in an abstract framework we refer to Appell (see [19]).

Note that when l.s.c. functionals of the form

$$F(u,v,B) = \int_B^* f(x,u,v)\, d\mu$$

are considered (without any measurability assumption on f), the integral representation Theorem 2.4.6 applies (the locality on $\mathfrak{I}$ and the additivity on $\mathfrak{I}$ being easy to prove), and we find the representation

$$F(u,v,B) = \int_B g(x,u,v)\, d\mu$$

where $g(x,s,z)$ is an actual normal-convex integrand.

Finally, we discuss briefly the case of operators between $L^p(\Omega;\mathbf{R}^m)$ and $L^q(\Omega;\mathbf{R}^n)$.

DEFINITION 2.7.1. *A mapping* $T:L^p(\Omega;\mathbf{R}^m)\to L^q(\Omega;\mathbf{R}^n)$ *is said to be a*

(i) **locally defined operator** *if it satisfies the following condition:*

for every $u,v\in L^p(\Omega;\mathbf{R}^m)$ *and* $B\in\mathfrak{I}$ *we have*

$$u=v \ \mu\text{-a.e. on } B \ \Rightarrow \ Tu=Tv \ \mu\text{-a.e. on } B;$$

(ii) **Nemyckii operator** *if there exist* n *integrands* $f_1,\ldots,f_n$ *such that*

$$(Tu)(x) = f(x,u(x))$$

where $f=(f_1,\ldots,f_n)$.

The following result about Nemyckii operators is well-known (see for instance

Vainberg [282]).

THEOREM 2.7.2. *Let* $f_1(x,s),\ldots,f_n(x,s)$ *be integrands, and let* $f=(f_1,\ldots,f_n)$. *Then, the following conditions are equivalent.*

(i) *the operator* $(T_f u)(x)=f(x,u(x))$ *is continuous from* $L^p(\Omega;\mathbf{R}^m)$ *into* $L^q(\Omega;\mathbf{R}^n)$;

(ii) $f_1,\ldots,f_n$ *are Carathéodory integrands and satisfy the following estimate for suitable* $a\in L^q(\Omega)$ *and* $b\geq 0$:

(2.5.1) $\quad |f(x,s)| \leq a(x) + b|s|^{p/q}$ *for* μ*-a.e.* $x\in\Omega$ *and for all* $s\in\mathbf{R}^m$.

It is clear that every Nemyckii operator is locally defined; the integral representation Theorem 2.4.6 allows us to invert in some cases this implication and to characterize all continuous and locally defined operators between $L^p(\Omega;\mathbf{R}^m)$ and $L^q(\Omega;\mathbf{R}^n)$.

THEOREM 2.7.3. *Let* $T:L^p(\Omega;\mathbf{R}^m)\to L^q(\Omega;\mathbf{R}^n)$ *be a locally defined operator which is continuous for the strong topologies of* $L^p(\Omega;\mathbf{R}^m)$ *and* $L^q(\Omega;\mathbf{R}^n)$. *Then, there exists a unique Carathéodory function* $f:\Omega\times\mathbf{R}^m\to\mathbf{R}^n$ *such that*

(i) $\quad |f(x,s)| \leq a(x) + b|s|^{p/q}$ *for* μ*-a.e.* $x\in\Omega$ *and all* $s\in\mathbf{R}^m$

(ii) $\quad (Tu)(x) = f(x,u(x))$ *for every* $u\in L^p(\Omega;\mathbf{R}^m)$ *and* μ*-a.e.* $x\in\Omega$

where a *is a suitable function in* $L^q(\Omega)$ *and* $b\geq 0$.

Proof. It is enough to consider the case $n=1$ and $q=1$. For every $u\in L^p(\Omega;\mathbf{R}^m)$ and $B\in\mathfrak{I}$ we set

$$F(u,B) = \int_B Tu \, d\mu \ .$$

The hypotheses on T imply that the functional F satisfies all conditions of Corollary

2.4.7, so that there exists a (unique) Carathéodory integrand $f(x,s)$ satisfying (i) and (ii). ■

REMARK 2.7.4. If we drop the continuity assumption on T in Theorem 2.7.3, the representation of T by means of a Nemyckii operator may fail. The operator T constructed in Section 2.5 is locally defined but, as shown in Theorem 2.5.1, it is not a Nemyckii operator. Nevertheless, it is possible to prove (see Appell [19]) that under the Continuum Hypothesis, for every locally defined operator T there exists a (possibly non measurable) function $f:\Omega\times\mathbf{R}^m\to\mathbf{R}^n$ such that for every $u\in L^p(\Omega;\mathbf{R}^m)$

(a) the map $x\to f(x,u(x))$ is measurable;

(b) $(Tu)(x)=f(x,u(x))$ a.e. on Ω.

CHAPTER 3

Functionals Defined on the Space of Measures

In this chapter we study the relaxation of functionals of the form

$$F(\lambda) = \begin{cases} \displaystyle\int_\Omega^* f(x,u)\, d\mu & \text{if } \lambda = u\cdot\mu \text{ with } u\in L^1(\Omega;\mathbf{R}^n) \\ +\infty & \text{otherwise,} \end{cases}$$

where λ belongs to the space $M(\Omega;\mathbf{R}^n)$ of vector-valued measures on Ω with bounded variation, $f:\Omega\times\mathbf{R}^n\to[0,+\infty]$ is a function, and $\int^*$ denotes the upper integral defined in Section 2.1. The topology we use is the weak* topology of measures. The proof of Theorem 3.3.1, which guarantees an integral representation for the relaxed functional, relies on an abstract integral representation theorem (Theorem 3.2.1) for functionals $F(\lambda,B)$ depending on measures $\lambda\in M(\Omega;\mathbf{R}^n)$ and Borel sets $B\in\mathbb{B}(\Omega)$.

In Section 3.4 we show some examples for which it is possible to compute explicitly the relaxed functional.

These results have been obtained by De Giorgi & Ambrosio & Buttazzo in [130] and by Ambrosio & Buttazzo in [14]; results of the same type, under measurability hypotheses on f, have been obtained in different frameworks and with different proofs by Bouchitte [57], Bouchitte & Valadier [58], Gavioli [165], Rockafellar [250], Valadier [283].

3.1. Notation and Preliminary Results

In all the chapter $(\Omega,\mathbb{B},\mu)$ will denote a measure space, where Ω is a separable locally compact metric space with distance d, $\mathbb{B}$ is the σ-algebra of Borel subsets of Ω, and $\mu:\mathbb{B}\to[0,+\infty[$ is a positive, finite, non-atomic measure. For every vector-valued measure $\lambda:\mathbb{B}\to\mathbf{R}^n$ and every $B\in\mathbb{B}$ the variation of λ on B is defined by

$$(3.1.1)\qquad |\lambda|(B) = \sup\Big\{\sum_{h=1}^{\infty}|\lambda(B_h)| \;:\; \bigcup_{h=1}^{\infty}B_h\subset B,\ B_h \text{ pairwise disjoint}\Big\}.$$

In this way, the set function $B\to|\lambda|(B)$ turns out to be a positive measure, which will be denoted by $|\lambda|$.

The following spaces will be considered.

$M(\Omega;\mathbf{R}^n)$ the space of all vector-valued measures $\lambda:\mathbb{B}\to\mathbf{R}^n$ with finite variation on Ω;

$C_c(\Omega;\mathbf{R}^n)$ the space of all continuous functions $u:\Omega\to\mathbf{R}^n$ with compact support;

$C_0(\Omega;\mathbf{R}^n)$ the space of all continuous functions $u:\Omega\to\mathbf{R}^n$ "vanishing on the boundary", that is for every $\varepsilon>0$ there exists a compact subset K_ε of Ω such that $|u(x)|<\varepsilon$ for all $x\in\Omega-K_\varepsilon$.

The space $C_0(\Omega;\mathbf{R}^n)$ endowed with the sup norm is a separable Banach space, and $C_c(\Omega;\mathbf{R}^n)$ is dense in $C_0(\Omega;\mathbf{R}^n)$. Moreover (see for instance Rudin [253], page 40), $M(\Omega;\mathbf{R}^n)$ can be identificated with the dual space of $C_0(\Omega;\mathbf{R}^n)$ by the duality

$$\langle\lambda,u\rangle_\Omega = \int_\Omega u\,d\lambda \qquad (\,u\in C_0(\Omega;\mathbf{R}^n),\ \lambda\in M(\Omega;\mathbf{R}^n)\,),$$

and for every $\lambda\in M(\Omega;\mathbf{R}^n)$ we have

$$(3.1.2)\qquad |\lambda|(\Omega) = \sup\Big\{|\langle\lambda,u\rangle_\Omega| \;:\; u\in C_0(\Omega;\mathbf{R}^n),\ \|u\|_{C_0(\Omega;\mathbf{R}^n)}\le 1\Big\}.$$

The space $M(\Omega;\mathbf{R}^n)$ will be endowed with the weak* topology deriving from the duality between $M(\Omega;\mathbf{R}^n)$ and $C_0(\Omega;\mathbf{R}^n)$; in particular, a sequence $\{\lambda_h\}$ in $M(\Omega;\mathbf{R}^n)$ will be said to w*-converge to $\lambda\in M(\Omega;\mathbf{R}^n)$ (and this will be indicated by $\lambda_h\to\lambda$) if and only if

$$\langle\lambda_h,u\rangle_\Omega \to \langle\lambda,u\rangle_\Omega \qquad \text{for every } u\in C_0(\Omega;\mathbf{R}^n)\,.$$

It is not difficult to see that $\lambda_h\to\lambda$ if and only if

$$\sup\{|\lambda_h|(\Omega) : h\in\mathbf{N}\} < +\infty \qquad \text{and} \qquad \langle\lambda_h,u\rangle\to\langle\lambda,u\rangle \quad \text{for every } u\in C_c(\Omega;\mathbf{R}^n).$$

Moreover, if $\lambda_h\in M(\Omega;\mathbf{R}^n)$ are positive measures and $\lambda_h\to\lambda$, then

$$\text{(3.1.3)} \qquad \int_\Omega u\, d\lambda \le \liminf_{h\to+\infty} \int_\Omega u\, d\lambda_h$$

for every non-negative l.s.c. function $u:\Omega\to\mathbf{R}^n$, and

$$\text{(3.1.4)} \qquad \int_C u\, d\lambda \ge \limsup_{h\to+\infty} \int_C u\, d\lambda_h$$

for every non-negative $u\in C_0(\Omega;\mathbf{R}^n)$ and every closed subset C of Ω.

DEFINITION 3.1.1. *We say that* $\lambda\in M(\Omega;\mathbf{R}^n)$ *is*

(i) **absolutely continuous** *with respect to* μ *(and we write* $\lambda<<\mu$ *) if*

$$|\lambda|(B)=0 \textit{ whenever } B\in\mathbb{B} \textit{ and } \mu(B)=0;$$

(ii) **singular** *with respect to* μ *(and we write* $\lambda\perp\mu$ *) if*

$$|\lambda|(\Omega-B)=0 \textit{ for a suitable } B\in\mathbb{B} \textit{ with } \mu(B)=0.$$

In the following, given $u\in L^1(\Omega;\mathbf{R}^n)$ we shall denote by $u\cdot\mu$ (or simply by u when no confusion is possible) the measure of $M(\Omega;\mathbf{R}^n)$ defined by

$$(u\cdot\mu)(B) = \int_B u\, d\mu \qquad \text{for every Borel set } B;$$

moreover, if $u:\Omega\to\mathbf{R}$ is a bounded Borel function and $\lambda\in M(\Omega;\mathbf{R}^n)$ we denote by $u\cdot\lambda$ the measure of $M(\Omega;\mathbf{R}^n)$

$$(u\cdot\lambda)(B) = \int_B u \, d\lambda \qquad \text{for every Borel set } B .$$

It is well-known that every absolutely continuous measure $\lambda\in M(\Omega;\mathbf{R}^n)$ is representable in the form $\lambda=u\cdot\mu$ for a suitable $u\in L^1(\Omega;\mathbf{R}^n)$; moreover, the following Lebesgue-Nikodym decomposition result for measures of $M(\Omega;\mathbf{R}^n)$ holds (see for instance Rudin [253], page 122).

PROPOSITION 3.1.2. *For every* $\lambda\in M(\Omega;\mathbf{R}^n)$ *there exists a unique function* $u\in L^1(\Omega;\mathbf{R}^n)$ *and a unique measure* $\lambda^s\in M(\Omega;\mathbf{R}^n)$ *such that*

(i) $\lambda = u\cdot\mu + \lambda^s$;

(ii) λ^s *is singular with respect to* μ.

The function u *is often indicated by* $\dfrac{d\lambda}{d\mu}$.

For every subset B of Ω we denote by 1_B the function

$$1_B(x) = \begin{cases} 1 & \text{if } x\in B \\ 0 & \text{if } x\in \Omega - B ; \end{cases}$$

moreover, for every $x\in\Omega$ we indicate by δ_x the measure of $M(\Omega;\mathbf{R}^n)$

$$\delta_x(B) = \begin{cases} 1 & \text{if } x\in B \\ 0 & \text{if } x\in \Omega - B . \end{cases}$$

For proper convex functions $f:\mathbf{R}^n\to]-\infty,+\infty]$ we define as usual the recession function of f (see Rockafellar [251]) by

$$f^{\infty}(z) = \lim_{t \to +\infty} \frac{f(w+tz)}{t} \qquad \text{for every } z \in \mathbf{R}^n ,$$

where w is any point in $\mathbf{R}^n$ such that $f(w)<+\infty$. In fact, the definition above is actually independent of the choice of w.

The following lemmas will be used.

LEMMA 3.1.3. *Let* $\{f_i\}_{i \in I}$ *be a family of convex l.s.c. functions from* $\mathbf{R}^n$ *into* $]-\infty,+\infty]$, *and let* $f=\sup\{f_i : i \in I\}$. *Assume* f *is proper. Then*

$$f^{\infty} = \sup \{f_i^{\infty} : i \in I\} .$$

Proof. Since $f_i \leq f$ for every $i \in I$, we obtain immediately

$$f_i^{\infty} \leq f^{\infty} \qquad \text{for every } i \in I,$$

so that

$$f^{\infty} \geq \sup \{f_i^{\infty} : i \in I\} . \tag{3.1.5}$$

Let us prove the opposite inequality. For every $i \in I$, $z \in \mathbf{R}^n$, $w \in \operatorname{dom} f_i$, $t>1$ we have

$$f_i(z+w) - f_i(w) = f_i\Big(\frac{1}{t}(tz+w) + (1-\frac{1}{t})w\Big) - f_i(w) \leq \frac{1}{t} f_i(tz+w) - \frac{1}{t} f_i(w)$$

hence, as $t \to +\infty$

$$f_i(z+w) - f_i(w) \leq f_i^{\infty}(z) . \tag{3.1.6}$$

Let now $z \in \mathbf{R}^n$, $w \in \operatorname{dom} f$, and $t>0$; we have $w \in \operatorname{dom} f_i$ for every $i \in I$, so that (3.1.6) yields

$$f(tz+w) - f(w) = \sup_{i \in I} f_i(tz+w) - \sup_{i \in I} f_i(w) \leq \sup_{i \in I} \big[f_i(tz+w) - f_i(w)\big] \leq$$

$$\leq \sup_{i\in I} f_i^\infty(tz) = t \sup_{i\in I} f_i^\infty(z) .$$

Therefore

$$f^\infty(z) = \lim_{t\to+\infty} \frac{f(tz+w) - f(w)}{t} \leq \sup_{i\in I} f_i^\infty(z)$$

which, together with (3.1.5), achieves the proof. ∎

LEMMA 3.1.4. *Let* $f:[0,+\infty[\to[0,+\infty]$ *be a convex function such that*

(i) $f(0) = 0$;

(ii) $f(2t) \leq 2f(t)$ *for every* $t\in[0,+\infty[$.

Then we have (with the convention $0\cdot\infty=0$ *)*

$$f(t) = t\, f(1) \qquad \textit{for every } t\in[0,+\infty[.$$

Proof. If $f(t)=+\infty$ for every $t>0$, the statement is obvious. If this is not the case, the function f is real valued, and making use of the convexity we find

$$f(2t) = 2\, f(t) \qquad \text{for every } t\in[0,+\infty[.$$

Now, the function

$$\varphi(t) = \frac{f(t)}{t}$$

is increasing in $]0,+\infty[$ and $\varphi(2t)=\varphi(t)$ for every $t\in]0,+\infty[$. Therefore φ is a constant function, and the lemma is proved. ∎

LEMMA 3.1.5. *Let* $f:\mathbf{R}^n\to]-\infty,+\infty]$ *be a proper convex function. Then, for every* $s,s_0\in\mathbf{R}^n$ *we have*

$$(3.1.7) \qquad f(s+s_0) \leq f(s_0) + f^\infty(s) .$$

Proof. By considering the function

$$g(s) = f(s+s_0)$$

we may reduce ourselves to prove the lemma for g and for $s_0=0$. If $g(0)=+\infty$ inequality (3.1.7) is trivial. If $g(0)<+\infty$, by convexity, for every $t>1$ we have

$$g(s) = g\Big(\frac{1}{t}\,ts + (1-\frac{1}{t})\,0\Big) \le \frac{g(ts)}{t} + (1-\frac{1}{t})\,g(0)\,,$$

and (3.1.7) follows by passing to the limit as $t\in +\infty$. ∎

We conclude this section with a result of convex analysis (see also Ekeland & Temam [148], page 14).

LEMMA 3.1.6. *Let* X *be a locally convex topological vector space and let* $F:X\to\,]-\infty,+\infty]$ *be a convex l.s.c. functional. Then, there exist two families* $\alpha_i\in\mathbf{R}$ *and* $\beta_i\in X'$ $(i\in I)$ *such that*

$$F(x) = \sup\{\alpha_i + \langle\beta_i,x\rangle \;:\; i\in I\} \qquad \textit{for every } x\in X.$$

Proof. If $F\equiv+\infty$ the assertion is obvious; hence we shall assume that there exists $z_0\in X$ with $F(z_0)<+\infty$. Since epi F is a closed convex set which does not contain the point $(z_0,F(z_0)-1)$, by the Hahn-Banach theorem there exist $\gamma\in X'$ and $m,n\in\mathbf{R}$ such that

$$\begin{cases} \langle\gamma,x\rangle + m\,y > n & \text{whenever } F(x)\le y \\ \langle\gamma,z_0\rangle + m\big(F(z_0)-1\big) < n\,. \end{cases}$$

Then, by taking $x= z_0$ and $y=F(z_0)$, we obtain $m>0$, so that

(3.1.8) $$F(x) > \frac{n}{m} - \langle\frac{\gamma}{m}, x\rangle \qquad \text{for every } x\in X.$$

In order to achieve the proof it will be enough to prove that for every $x_0 \in X$ and every $a<F(x_0)$ there exist $\alpha \in \mathbf{R}$ and $\beta \in X'$ such that

$$(3.1.9) \qquad \begin{cases} F(x) \geq \alpha + \langle \beta, x \rangle & \text{for every } x \in X \\ a \leq \alpha + \langle \beta, x_0 \rangle . \end{cases}$$

Let us fix $x_0 \in X$ and $a<F(x_0)$; by the Hahn-Banach theorem again, there exist $\delta \in X'$ and $p,q \in \mathbf{R}$ such that

$$(3.1.10) \qquad \begin{cases} \langle \delta, x \rangle + p\,y > q & \text{whenever } F(x) \leq y \\ \langle \delta, x_0 \rangle + p\,a < q . \end{cases}$$

Since $\operatorname{dom} F = \{x \in X : F(x) < +\infty\}$ is not empty, from (3.1.10) we have $p \geq 0$. Assume $p>0$; by (3.1.10) we obtain

$$F(x) > \frac{q}{p} - \langle \frac{\delta}{p}, x \rangle \qquad \text{for every } x \in X$$

$$a < \frac{q}{p} - \langle \frac{\delta}{p}, x_0 \rangle ,$$

so that (3.1.9) is proved with $\alpha = {}^q/_p$ and $\beta = -{}^\delta/_p$. Assume now $p=0$; then (3.1.10) becomes

$$\langle \delta, x_0 \rangle < q < \langle \delta, x \rangle \qquad \text{for every } x \in \operatorname{dom} F$$

and, taking into account (3.1.8), we obtain for every $c>0$

$$F(x) > \frac{n}{m} - \langle \frac{\gamma}{m}, x \rangle + c\big(q - \langle \delta, x \rangle\big) .$$

Finally, since $\langle \delta, x_0 \rangle < q$, by choosing c large enough we obtain also

$$a < \frac{n}{m} - \langle \frac{\gamma}{m}, x_0 \rangle + c\big(q - \langle \delta, x_0 \rangle\big)$$

which proves (3.1.9) with $\alpha = cq + {}^n/_m$ and $\beta = -c\delta - {}^\gamma/_m$. ∎

3.2. Integral Representation

In this section we prove the following theorem.

THEOREM 3.2.1. *Let* $F:M(\Omega;\mathbf{R}^n)\times\mathbb{B}\to]-\infty,+\infty]$ *be a functional satisfying the following conditions:*

(i) F *is* $\mathbb{B}$*-local, that is*

$$\lambda,\nu\in M(\Omega;\mathbf{R}^n),\ B\in\mathbb{B},\ |\lambda-\nu|(B)=0 \quad\Rightarrow\quad F(\lambda,B)=F(\nu,B)\,;$$

(ii) *for every* $\lambda\in M(\Omega;\mathbf{R}^n)$ *the set function* $F(\lambda,\cdot)$ *is additive;*

(iii) *the function* $F(\cdot,\Omega)$ *is convex and seq.* $w^*M(\Omega;\mathbf{R}^n)$*-l.s.c.;*

(iv) *there exists* $u_0\in L^1(\Omega;\mathbf{R}^n)$ *such that* $F(u_0,B)<+\infty$ *for every* $B\in\mathbb{B}$;

(v) *for every* $\lambda\in M(\Omega;\mathbf{R}^n)$ *with* $\lambda\perp\mu$ *the function* $t\to F(t\lambda+u_0,\Omega)-F(u_0,\Omega)$ *is positively* 1*-homogeneous.*

Then, there exists a Borel function $f:\Omega\times\mathbf{R}^n\to]-\infty,+\infty]$ *such that*

(a) *for* μ*-a.e.* $x\in\Omega$ *the function* $f(x,\cdot)$ *is convex and l.s.c. on* $\mathbf{R}^n$;

(b) *there exist* $a\in L^1(\Omega)$ *and* $b\geq 0$ *with*

$$f(x,s)\ \geq\ -\left[a(x)+b|s|\right] \qquad \text{for } \mu\text{-a.e. } x\in\Omega \text{ and for all } s\in\mathbf{R}^n;$$

(c) *for every* $\lambda\in M(\Omega;\mathbf{R}^n)$ *and every* $B\in\mathbb{B}$

$$F(\lambda,B)-F(u_0,B) = \int_B f(x,\frac{d\lambda}{d\mu})\,d\mu + \int_B f^\infty(x,\frac{d\lambda^s}{d|\lambda|})\,d|\lambda|$$

where $\lambda = {}^{d\lambda}/_{d\mu}\cdot\mu+\lambda^s$ *is the Lebesgue-Nikodym decomposition of* λ, *and for every* $x\in\Omega$ $f^\infty(x,\cdot)$ *is the recession function of* $f(x,\cdot)$.

(d) *the function* $f^\infty(x,s)$ *is l.s.c. in* (x,s).

We begin by proving some preliminary results.

PROPOSITION 3.2.2. *Assume* F *satisfies all conditions (i),...,(v) of Theorem 3.2.1. Then, there exist* $a \in L^1(\Omega)$ *and* $b \geq 0$ *such that for every* $\lambda \in M(\Omega;\mathbf{R}^n)$ *and* $B \in \mathbb{B}$

$$F(\lambda,B) - F(u_0,B) \geq -\Big[\int_B a\,d\mu + b\,|\lambda|(B)\Big] .$$

Proof. By considering the functional

$$G(\lambda,B) = F(\lambda+u_0,B) - F(u_0,B) \qquad (\lambda \in M(\Omega;\mathbf{R}^n) \text{ and } B \in \mathbb{B})$$

we may limit ourselves to the case $u_0=0$ and $F(0,B)=0$ for every $B \in \mathbb{B}$. By Lemma 2.4.5 there exist $a \in L^1(\Omega)$ and $b_1 \geq 0$ such that

$$(3.2.1) \qquad F(u,B) \geq -\int_B \big[a(x)+b_1|u|\big]\,d\mu$$

for every $u \in L^1(\Omega;\mathbf{R}^n)$ and $B \in \mathbb{B}$. We claim that for a suitable constant $b_2 \geq 0$ we have

$$(3.2.2) \qquad F(\lambda,\Omega) \geq -b_2\,|\lambda|(\Omega)$$

for every $\lambda \in M(\Omega;\mathbf{R}^n)$ with $\lambda \perp \mu$. In fact, if (3.2.2) is not satisfied, by using assumption (v) of Theorem 3.2.1 we can find a sequence $\{\lambda_h\}$ in $M(\Omega;\mathbf{R}^n)$ with $\lambda_h \perp \mu$ for every $h \in \mathbf{N}$ and such that

$$|\lambda_h|(\Omega) = 1 \qquad \text{and} \qquad F(\lambda_h,\Omega) \leq -h \qquad \text{for every } h \in \mathbf{N} .$$

By extracting a subsequence $\{\lambda_{h(k)}\}$ which $w^*M(\Omega;\mathbf{R}^n)$-converges to some $\lambda \in M(\Omega;\mathbf{R}^n)$ we obtain by lower semicontinuity

$$F(\lambda,\Omega) \leq \liminf_{k\to+\infty} F(\lambda_{h(k)},\Omega) = -\infty ,$$

which contradicts the fact that F takes its values in $]-\infty,+\infty]$.

Let now $b=\max\{b_1,b_2\}$; by Proposition 3.1.2 for every $\lambda \in M(\Omega;\mathbf{R}^n)$ we have

$\lambda=u\cdot\mu+\lambda^s$ for suitable $u\in L^1(\Omega;\mathbf{R}^n)$ and $\lambda^s\perp\mu$. Then, by using the locality and the additivity of F we have from (3.2.1) and (3.2.2) that for every $B\in\mathbb{B}$

$$F(\lambda,B) = F(u,B) + F(\lambda^s,B) \geq$$

$$\geq -b\left[\int_B |u|\,d\mu + |\lambda^s|(B)\right] - \int_B a\,d\mu = -\left[\int_B a\,d\mu + b\,|\lambda|(B)\right] . \blacksquare$$

PROPOSITION 3.2.3. *Assume* F *satisfies all conditions (i),...,(v) of Theorem 3.2.1. Then, for every* $\lambda\in M(\Omega;\mathbf{R}^n)$ *the set function* $B\to F(\lambda,B)-F(u_0,B)$ *is σ-additive on* $\mathbb{B}$.

Proof. As in the proof of Proposition 3.2.2 we may assume $u_0=0$ and $F(0,B)=0$ for every $B\in\mathbb{B}$. Let $\lambda\in M(\Omega;\mathbf{R}^n)$ be fixed; it is enough to prove that for every sequence $\{B_h\}$ in $\mathbb{B}$ with $B_h\uparrow B$ we have

$$F(\lambda,B) = \lim_{h\to+\infty} F(\lambda,B_h) . \tag{3.2.3}$$

Let $B_h\uparrow B$; since $1_{B_h}\cdot\mu\to 1_B\cdot\mu$ in $w^*M(\Omega;\mathbf{R}^n)$, by the lower semicontinuity and the locality of F we obtain

$$F(\lambda,B) \leq \liminf_{h\to+\infty} F(\lambda,B_h) . \tag{3.2.4}$$

On the other hand, Proposition 3.2.2 entails that there exist $a\in L^1(\Omega)$ and $b\geq 0$ with

$$F(\lambda,A) + b\,|\lambda|(A) + \int_A a\,d\mu \geq 0$$

for every $A\in\mathbb{B}$; therefore

$$\limsup_{h\to+\infty} \left[F(\lambda,B_h) + b|\lambda|(B_h) + \int_{B_h} a\,d\mu\right] \leq F(\lambda,B) + b|\lambda|(B) + \int_B a\,d\mu ,$$

so that

$$\limsup_{h\to+\infty} F(\lambda,B_h) \le F(\lambda,B)$$

which, together with (3.2.4), proves (3.2.3). ■

PROPOSITION 3.2.4. *Assume* F *satisfies all conditions (i),...,(v) of Theorem 3.2.1. and the additional conditions:*

(vi) $F(0,B)=0$ *for every* $B\in\mathbb{B}$;

(vii) *there exists* $k\ge 0$ *such that*

$$|F(\lambda,B) - F(\nu,B)| \le k\,|\lambda-\nu|(B) \quad \textit{for every } \lambda,\nu\in M(\Omega;\mathbf{R}^n) \textit{ and } B\in\mathbb{B}.$$

Then, there exists a Borel function $f:\Omega\times\mathbf{R}^n\to\mathbf{R}$ *such that*

(a) *for* μ*-a.e.* $x\in\Omega$ *the function* $f(x,\cdot)$ *is convex and k-Lipschitz continuous;*

(b) $f^\infty(x,s)$ *is l.s.c. in* (x,s);

(c) *for every* $\lambda\in M(\Omega;\mathbf{R}^n)$ *and* $B\in\mathbb{B}$

$$F(\lambda,B) = \int_B f(x,\frac{d\lambda}{d\mu})\,d\mu + \int_B f^\infty(x,\frac{d\lambda^s}{d|\lambda|})\,d|\lambda| .$$

Proof. The functional defined on $L^1(\Omega;\mathbf{R}^n)\times\mathbb{B}$ by

$$G(u,B) = F(u\cdot\mu,B)$$

satisfies all conditions of the integral representation Theorem 2.4.6, so it can be written in the form

$$G(u,B) = \int_B g(x,u)\,d\mu \tag{3.2.5}$$

for every $u\in L^1(\Omega;\mathbf{R}^n)$ and $B\in\mathbb{B}$, where $g(x,s)$ is a suitable convex integrand which can be assumed to be a Borel function. From conditions (vi) and (vii) we easily see that it is not restrictive to assume

(3.2.6) for every $x \in \Omega$ $g(x,0)=0$;

(3.2.7) for every $x \in \Omega$ the function $g(x,\cdot)$ is convex;

(3.2.8) for every $x \in \Omega$ and $s_1,s_2 \in \mathbf{R}^n$ $|g(x,s_1)-g(x,s_2)| \leq k\, |s_1-s_2|$.

Let $\psi:\Omega\times\mathbf{R}^n\to\mathbf{R}$ be the function defined by

$$\psi(x,s) = F(s\cdot\delta_x,\Omega)\,. \tag{3.2.9}$$

The lower semicontinuity of F entails that ψ is l.s.c. in $\Omega\times\mathbf{R}^n$, and from hypothesis (v) of Theorem 3.2.1 we get

$$F(t\lambda,\Omega) = t\,F(\lambda,\Omega)$$

for every $t\geq 0$ and every $\lambda\in M(\Omega;\mathbf{R}^n)$ with $\lambda\perp\mu$. Therefore $\psi(x,s)$ is positively 1-homogeneous and (by condition (vii)) Lipschitz continuous in s. We shall show that

$$F(\lambda,B) = \int_B \psi\Big(x,\frac{d\lambda}{d|\lambda|}\Big)\, d|\lambda| \tag{3.2.10}$$

for every $B\in\mathbb{B}$ and every $\lambda\in M(\Omega;\mathbf{R}^n)$ with $\lambda\perp\mu$. Let us fix $\lambda\in M(\Omega;\mathbf{R}^n)$ with $\lambda\perp\mu$; since F is $\mathbb{B}$-local, by (vi) and (vii) the measure $F(\lambda,\cdot)$ is absolutely continuous with respect to $|\lambda|$, so there exists $\theta\in L^1(\Omega,|\lambda|)$ such that for every $B\in\mathbb{B}$

$$F(\lambda,B) = \int_B \theta\, d|\lambda|\,.$$

Denote by w the function $d\lambda/d|\lambda|$; in order to prove (3.2.10) we have to show that

$$\theta(x) = \psi(x,w(x)) \qquad \text{for } |\lambda|\text{-a.e. } x\in\Omega\,.$$

By using Scorza Dragoni theorem (see for instance Ekeland & Temam [148], page 235) for every $\varepsilon>0$ we may find a compact subset K_ε of Ω such that $\mu(K_\varepsilon)=0$, $|\lambda|(\Omega-K_\varepsilon)<\varepsilon$, and the functions $w:K_\varepsilon\to\mathbf{R}^n$, $\theta:K_\varepsilon\to\mathbf{R}^n$, $\psi:K_\varepsilon\times\mathbf{R}^n\to\mathbf{R}$ are continuous. Let us denote by T_ε the set

$$T_\varepsilon = \{x\in K_\varepsilon : |\lambda|(U\cap K_\varepsilon)=0 \text{ for some open set } U \text{ containing } x\}\,;$$

T_ε is a Borel set and, since Ω is separable, we have $|\lambda|(T_\varepsilon)=0$. Let $y\in K_\varepsilon - T_\varepsilon$ and set for every $i\in\mathbf{N}$

$$C_i = \{x\in\Omega : d(x,y)<2^{-i}\} \qquad c_i = |\lambda|(C_i) \qquad \nu_i = 1_{C_i}\cdot\lambda \ .$$

Since $w:K_\varepsilon\to\mathbf{R}^n$ is continuous at y, we have as $i\to+\infty$

$$\frac{1}{c_i}\nu_i \to w(y)\cdot\delta_y \qquad \text{in the topology } w^*M(\Omega;\mathbf{R}^n)\,,$$

therefore, making use of the continuity of $\theta:K_\varepsilon\to\mathbf{R}^n$ at y, we obtain

$$\psi(y,w(y)) = F(w(y)\cdot\delta_y,\Omega) \le \liminf_{i\to+\infty} F(\frac{1}{c_i}\cdot\nu_i,\Omega) =$$

$$= \liminf_{i\to+\infty} \int_{C_i} \frac{\theta}{c_i}\, d|\lambda| = \theta(y)\ ,$$

so that

(3.2.11) $$\psi(x,w(x)) \le \theta(x) \qquad \text{for every } x\in K_\varepsilon - T_\varepsilon\,.$$

In order to prove the opposite inequality, for every $h\in\mathbf{N}$ let $\{C_{i,h}\}_{1\le i\le k(h)}$ be a finite partition of K_ε into Borel sets having diameter less than 2^{-h}; if we choose $x_{i,h}\in C_{i,h}$ it is not difficult to see that as $h\to+\infty$

$$\lambda_h = \sum_{i=1}^{k(h)} \lambda(C_{i,h})\cdot\delta_{x_{i,h}} \to 1_{K_\varepsilon}\cdot\lambda \qquad \text{in the topology } w^*M(\Omega;\mathbf{R}^n)\,.$$

Recalling hypothesis (iii) and (vii) we get

$$\int_{K_\varepsilon} \theta\, d|\lambda| = F(\lambda,K_\varepsilon) \le \liminf_{h\to+\infty} F(\lambda_h,K_\varepsilon) =$$

$$= \liminf_{h\to+\infty} \sum_{i=1}^{k(h)} \psi(x_{i,h},\lambda(C_{i,h})) \le$$

$$\le \liminf_{h\to+\infty} \sum_{i=1}^{k(h)} \Big\{ \psi(x_{i,h},w(x_{i,h}))\,|\lambda|(C_{i,h}) + k\,\Big| w(x_{i,h})\,|\lambda|(C_{i,h}) - \int_{C_{i,h}} w\, d|\lambda| \Big| \Big\} =$$

$$= \liminf_{h\to+\infty} \sum_{i=1}^{k(h)} \psi(x_{i,h},w(x_{i,h}))\ |\lambda|(C_{i,h}) = \int_{K_\varepsilon} \psi(x,w)\, d|\lambda| ,$$

which, together with (3.2.11) concludes the proof of (3.2.10).

Let now $B\in\mathbb{B}$ and $\lambda\in M(\Omega;\mathbf{R}^n)$; denoting by λ^s the singular part of λ with respect to μ and by M a Borel set such that $\mu(M)=|\lambda^s|(\Omega-M)=0$, we have

(3.2.12) $$F(\lambda,B) = F(\lambda,B-M) + F(\lambda,B\cap M) =$$

$$= F(\frac{d\lambda}{d\mu}\cdot\mu,B-M) + F(\lambda^s,B\cap M) = \int_{B-M} g(x,\frac{d\lambda}{d\mu})\, d\mu + \int_{B\cap M} \psi(x,\frac{d\lambda^s}{d|\lambda|})\, d|\lambda| =$$

$$= \int_B g(x,\frac{d\lambda}{d\mu})\, d\mu + \int_B \psi(x,\frac{d\lambda^s}{d|\lambda|})\, d|\lambda| .$$

In order to conclude the proof, we show that

(3.2.13) $$g^\infty(x,s) = \psi(x,s) \quad \text{for } \mu\text{-a.e. } x\in\Omega \text{ and for all } s\in\mathbf{R}^n.$$

Since $g(x,s)$ and $\psi(x,s)$ are Lipschitz continuous in s, it will be enough to show that

$$g^\infty(x,s) = \psi(x,s) \quad \text{for } \mu\text{-a.e. } x\in\Omega \text{ and for all } s\in\mathbf{Q}^n.$$

Fix $s\in\mathbf{Q}^n$ and set

$$E = \{x\in\Omega : \mu(U)=0 \text{ for some open set } U \text{ containing } x\}.$$

The set E is open and $\mu(E)=0$; moreover, setting for every $y\in\Omega-E$ and $i\in\mathbf{N}$

$$B_i = \{x\in\Omega : d(x,y)<2^{-i}\} \qquad \text{and} \qquad c_i = \mu(B_i)$$

we have as $i\to+\infty$

$$\frac{s}{c_i} 1_{B_i}\cdot\mu \to s\cdot\delta_y \qquad \text{in the topology } w^*M(\Omega;\mathbf{R}^n) ,$$

therefore

$$\psi(y,s) = F(s\cdot\delta_y,\Omega) \le \liminf_{i\to+\infty} F(\frac{s}{c_i} 1_{B_i}\cdot\mu,\Omega) =$$

$$= \liminf_{i\to+\infty} \frac{1}{c_i} \int_{B_i} c_i\, g(x,\frac{s}{c_i})\, d\mu \le \liminf_{i\to+\infty} \frac{1}{c_i} \int_{B_i} g^\infty(x,s)\, d\mu .$$

From the Lebesgue derivation theorem (see for instance Federerer [151], Theorem 2.2.9) we get

$$\psi(x,s) \le g^\infty(x,s) \quad \text{for } \mu\text{-a.e. } x\in\Omega.$$

In order to obtain the opposite inequality, since for every $x\in\Omega$ the function $\psi(x,\cdot)$ is positively 1-homogeneous and

$$g^\infty(x,s) = \sup\left\{\frac{g(x,ts)}{t} : t>0\right\} ,$$

we only need to show that

$$g(x,s) \le \psi(x,s) \quad \text{for } \mu\text{-a.e. } x\in\Omega. \tag{3.2.14}$$

By contradiction, if inequality (3.2.14) is false, by Lusin theorem we can find a compact subset K of Ω and a number $\varepsilon>0$ such that

$$\begin{cases} \mu(K) > 0 \\ g(x,s) \ge \psi(x,s) + \varepsilon \quad \text{for every } x\in K \\ \psi(\cdot,s):K\to\mathbf{R} \text{ is continuous} . \end{cases}$$

For every $h\in\mathbf{N}$ let $\{C_{i,h}\}_{1\le i\le k(h)}$ be a finite partition of K into Borel sets having diameter less than 2^{-h}; if we choose $x_{i,h}\in C_{i,h}$ we have as $h\to+\infty$

$$\lambda_h = \sum_{i=1}^{k(h)} s\,\mu(C_{i,h})\cdot\delta_{x_{i,h}} \to s\,1_K\cdot\mu \qquad \text{in the topology } w^*M(\Omega;\mathbf{R}^n),$$

therefore

$$\int_K g(x,s)\, d\mu = F(s\,1_K\cdot\mu,\Omega) \le$$

$$\le \liminf_{h\to+\infty} F(\lambda_h,\Omega) = \liminf_{h\to+\infty} \sum_{i=1}^{k(h)} \int_{C_{i,h}} \psi(x_{i,h},s)\, d\mu =$$

$$= \int_K \psi(x,s)\, d\mu \;\le\; \int_K g(x,s)\, d\mu - \varepsilon\, \mu(K)$$

which is a contradiction. Define now

$$f(x,s) \;=\; \begin{cases} g(x,s) & \text{if } g^\infty(x,s)=\psi(x,s) \\ \psi(x,s) & \text{otherwise}\,; \end{cases}$$

we have $f^\infty(x,s)=\psi(x,s)$ for every $(x,s)\in \Omega\times\mathbf{R}^n$, hence $f^\infty(x,s)$ is l.s.c. in (x,s), and by (3.2.12) and (3.2.13)

$$F(\lambda,B) \;=\; \int_B f(x,\frac{d\lambda}{d\mu})\, d\mu \;+\; \int_B f^\infty(x,\frac{d\lambda^s}{d|\lambda|})\, d|\lambda|$$

for all $\lambda\in M(\Omega;\mathbf{R}^n)$ and $B\in\mathbb{B}$. ■

Proof of Theorem 3.2.1. As we already noticed, there is no loss of generality if we assume $u_0=0$ and $F(0,B)=0$ for every $B\in\mathbb{B}$. Set for every $\lambda\in M(\Omega;\mathbf{R}^n)$ and $B\in\mathbb{B}$

$$G(\lambda,B) \;=\; F(\lambda,B) + \int_B a\, d\mu \;+ b\,|\lambda|(B) \tag{3.2.15}$$

where $a\in L^1(\Omega)$ and $b\ge 0$ are given by Proposition 3.2.2, and for every $k\in\mathbf{N}$

$$G_k(\lambda,B) \;=\; \inf\left\{ G(v,B) + |\lambda-v|(B) \;:\; v\in M(\Omega;\mathbf{R}^n)\right\}\,.$$

Since $G\ge 0$ and $G(0,B)=\int_B a\, d\mu$, we have

$$0 \;\le\; G_k(0,B) \;\le\; \int_B a\, d\mu$$

for every $k\in\mathbf{N}$ and $B\in\mathbb{B}$. Then, the measures $G_k(0,\cdot)$ are absolutely continuous with respect to μ, so that

$$G_k(0,B) = \int_B a_k \, d\mu$$

for suitable $a_k \in L^1(\Omega)$. Moreover, by Proposition 1.3.7, we get as $k \to +\infty$

(3.2.16) $\qquad G_k(\lambda,B) \uparrow G(\lambda,B) \qquad$ for every $\lambda \in M(\Omega;\mathbf{R}^n)$ and $B \in \mathbb{B}$.

We want to apply Proposition 3.2.4 to the functionals $G_k(\lambda,B) - G_k(0,B)$. Properties (i), (ii), (iv), (vi), (vii) are easy to prove; we prove now property (iii). From the definition of G_k and the convexity of G we get immediately that $G_k(\cdot,\Omega)$ is convex. Let $\lambda_h \to \lambda$ in $w^*M(\Omega;\mathbf{R}^n)$; it is not restrictive to assume that $G_k(\lambda_h,\Omega)$ converges (as $h \to +\infty$) to a finite limit. Moreover, in the definition of G_k the infimum is actually attained, so that for every $h \in \mathbf{N}$ there exists $v_h \in M(\Omega;\mathbf{R}^n)$ with

(3.2.17) $\qquad G_k(\lambda_h,\Omega) = G(v_h,\Omega) + k\,|\lambda_h - v_h|(\Omega)$.

It is easy to see that the sequence $\{v_h\}$ is bounded in $M(\Omega;\mathbf{R}^n)$, and we may assume it converges in $w^*M(\Omega;\mathbf{R}^n)$ to some $v \in M(\Omega;\mathbf{R}^n)$. Then, by (3.2.17) we have

$$G_k(\lambda,\Omega) \le G(v,\Omega) + k\,|\lambda - v|(\Omega) \le$$

$$\le \liminf_{h\to+\infty} \Big[G(v_h,\Omega) + k\,|\lambda_h - v_h|(\Omega) \Big] = \liminf_{h\to+\infty} G_k(\lambda_h,\Omega) \ .$$

Finally, we prove property (v). Let $t>0$, $\lambda \in M(\Omega;\mathbf{R}^n)$ with $\lambda \perp \mu$, and $B \in \mathbb{B}$; in the case $\mu(B)=0$ we have

$$G_k(t\lambda,B) = \inf\Big\{ G(v,B) + k\,|t\lambda - v|(B) \ : \ v \in M(\Omega;\mathbf{R}^n) \Big\} =$$

$$= \inf\Big\{ G(tv,B) + tk\,|\lambda - v|(B) \ : \ v \in M(\Omega;\mathbf{R}^n) \Big\} =$$

$$= t \inf\Big\{ G(v,B) + k\,|\lambda - v|(B) \ : \ v \in M(\Omega;\mathbf{R}^n) \Big\} = t\, G_k(\lambda,B) \ .$$

In the general case $B \in \mathbb{B}$, if we denote by M a μ-negligible set in $\mathbb{B}$ such that $|\lambda|(\Omega - M) = 0$, by the locality and additivity properties we get

$$G_k(t\lambda,B) - G_k(0,B) = G_k(t\lambda,B \cap M) + G_k(t\lambda,B - M) - G_k(0,B) =$$

$$= t\,G_k(\lambda,B\cap M) \;=\; t\left[G_k(\lambda,B) - G_k(\lambda,B-M)\right] \;=$$

$$= t\left[G_k(\lambda,B) - G_k(0,B)\right] .$$

Then, Proposition 3.2.4 applies, and we obtain for every $\lambda\in M(\Omega;\mathbf{R}^n)$ and $B\in\mathbb{B}$

$$G_k(\lambda,B) \;=\; \int_B g_k(x,\frac{d\lambda}{d\mu})\,d\mu \;+\; \int_B g_k^\infty(x,\frac{d\lambda^s}{d|\lambda|})\,d|\lambda| \tag{3.2.18}$$

for a suitable Borel function $g_k(x,s)$ convex in s and with $g_k(x,s)$ l.s.c. in (x,s). By Proposition 2.1.3 there exists $N\in\mathbb{B}$ with $\mu(N)=0$ such that

$$0 \le g_k(x,s) \le g_m(x,s) \qquad \text{for every } x\in\Omega-N \text{ and } s\in\mathbf{R}^n$$

whenever $k,m\in\mathbf{N}$ with $k\le m$. Since $g_k^\infty(x,s)=G_k(s\cdot\delta_x,\{x\})$, we have

$$0 \le g_k^\infty(x,s) \le g_m^\infty(x,s) \qquad \text{for every } x\in\Omega \text{ and } s\in\mathbf{R}^n$$

whenever $k,m\in\mathbf{N}$ with $k\le m$. Set now for every $x\in\Omega$ and $s\in\mathbf{R}^n$

$$g(x,s) \;=\; \begin{cases} \sup\,\{g_k(x,s) : k\in\mathbf{N}\} & \text{if } x\in\Omega-N \\ \sup\,\{g_k^\infty(x,s) : k\in\mathbf{N}\} & \text{if } x\in N\,. \end{cases}$$

By using Lemma 3.1.3 formulas (3.2.16) and (3.2.18) yield immediately that

$$G(\lambda,B) \;=\; \int_B g(x,\frac{d\lambda}{d\mu})\,d\mu \;+\; \int_B g^\infty(x,\frac{d\lambda^s}{d|\lambda|})\,d|\lambda|$$

for every $\lambda\in M(\Omega;\mathbf{R}^n)$ and $B\in\mathbb{B}$. Moreover, g is a convex integrand and $g^\infty(x,s)$ is l.s.c. in (x,s). Finally, setting

$$f(x,s) \;=\; g(x,s) - b\,|s| - a(x)$$

by (3.2.15) we get

$$F(\lambda,B) \;=\; \int_B f(x,\frac{d\lambda}{d\mu})\,d\mu \;+\; \int_B f^\infty(x,\frac{d\lambda^s}{d|\lambda|})\,d|\lambda|$$

for every $\lambda\in M(\Omega;\mathbf{R}^n)$ and $B\in\mathbb{B}$. Since the functional

$$u \to \int_B f(x,u)\, d\mu = F(u\cdot\mu,B)$$

is sequentially weakly l.s.c. in $L^1(\Omega;\mathbf{R}^n)$, the convexity of $f(x,\cdot)$ follows from Theorem 2.3.6 and Remark 2.3.7. ■

3.3. Relaxation in the Space of Measures

Let $f:\Omega\times\mathbf{R}^n\to[0,+\infty]$ be a given function (note that no measurability hypotheses are required); for every $\lambda\in M(\Omega;\mathbf{R}^n)$ and $B\in\mathbb{B}$ define

$$(3.3.1) \qquad F(\lambda,B) = \begin{cases} \displaystyle\int_B^* f(x,u)\, d\mu & \text{if } \lambda=u\cdot\mu \text{ with } u\in L^1(\Omega;\mathbf{R}^n) \\ +\infty & \text{otherwise} \end{cases}$$

where $\int^*$ denotes the upper integral defined in Section 2.1. In this section we shall prove that the relaxed functional

$$\Gamma F(\lambda,B) = \Gamma_{seq}(w^*M(\Omega;\mathbf{R}^n)^-)\, F(\lambda,B)$$

can be written in the integral form

$$(3.3.2) \qquad \Gamma F(\lambda,B) = \int_B \varphi(x,\frac{d\lambda}{d\mu})\, d\mu + \int_B \varphi^\infty(x,\frac{d\lambda^s}{d|\lambda|})\, d|\lambda|$$

for a suitable function φ. More precisely, the following theorem holds.

THEOREM 3.3.1. *Assume the functional* $F(\cdot,\Omega)$ *defined in (3.3.1) is finite in at least one* $u_0\in L^1(\Omega;\mathbf{R}^n)$. *Then, there exists a Borel function* $\varphi:\Omega\times\mathbf{R}^n\to[0,+\infty]$ *such that*

(a) *for* μ*-a.e.* $x\in\Omega$ *the function* $\varphi(x,\cdot)$ *is convex and l.s.c. on* $\mathbf{R}^n$;

(b) *formula (3.3.2) holds for every* $\lambda\in M(\Omega;\mathbf{R}^n)$ *and every* $B\in\mathbb{B}$;

(c) *the recession function* $\varphi^\infty(x,s)$ *is l.s.c. in* (x,s).

REMARK 3.3.2. Let us denote by $G(u,B)$ the greatest functional which is sequentially weakly l.s.c. in $L^1(\Omega;\mathbf{R}^n)$ and less than or equal to $F(u,B)$. It is easy to see that the functional G satisfies all hypotheses of Theorem 2.4.2, so that

$$G(u,B) = \int_B g(x,u)\, d\mu$$

for a suitable convex integrand g. Moreover, the relaxed functional ΓF can be seen as the greatest functional which is sequentially $w^*M(\Omega;\mathbf{R}^n)$-l.s.c. and less than or equal to G on $L^1(\Omega;\mathbf{R}^n)$. Thus, in the following we may assume that f is a convex integrand, so that the upper integral becomes an usual integral and the functional $F(\cdot,B)$ may be assumed to be convex for every $B\in\mathbb{B}$.

Let $\mathbb{A}$ be the family of all open subsets of Ω; given $A\in\mathbb{A}$ and $\lambda_h,\lambda\in M(\Omega;\mathbf{R}^n)$ $(h\in\mathbf{N})$, we say that $\lambda_h\to\lambda$ in A if for every $u\in C_0(A;\mathbf{R}^n)$ we have

$$\lim_{h\to+\infty}\int_A u\, d\lambda_h = \int_A u\, d\lambda \ .$$

LEMMA 3.3.3. *Let* $\lambda_h,\lambda\in M(\Omega;\mathbf{R}^n)$ *and let* $\nu\in M(\Omega;\mathbf{R}^n)$, $A\in\mathbb{A}$ *be such that*

$$\lambda_h\to\lambda \text{ in } A\,, \qquad |\lambda_h|\to\nu\,, \qquad \nu(\partial A)=0 .$$

Then we have $1_A\cdot\lambda_h\to 1_A\cdot\lambda$.

Proof. For every $\rho>0$ set

$$A_\rho = \{x\in\Omega : d(x,\Omega-A)\geq\rho\}$$
$$B_\rho = \{x\in\Omega : d(x,\Omega-A)\leq\rho\}$$
$$C_\rho = B_\rho\cap\overline{A}$$

and let $\varphi_\rho:\Omega\to\mathbf{R}$ be a continuous function such that

$$0\leq\varphi_\rho\leq 1\,, \qquad \varphi_\rho=1 \text{ on } A_{2\rho}\,, \qquad \varphi_\rho=0 \text{ on } B_\rho\,.$$

Fix $u\in C_0(\Omega;\mathbf{R}^n)$; by using (3.1.4) we get

$$\limsup_{h\to+\infty}\left|\int_A u\varphi_\rho\,d\lambda_h - \int_A u\,d\lambda_h\right| \leq$$
$$\leq \limsup_{h\to+\infty}\int_{C_{2\rho}}|u|\,d|\lambda_h| \leq \int_{C_{2\rho}}|u|\,d\nu$$

so that, since $\nu(A)=0$, $C_\rho\downarrow\partial A$ and $u\varphi_\rho\in C_0(\Omega;\mathbf{R}^n)$

$$\int_A u\,d\lambda = \lim_{\rho\to 0}\int_A u\varphi_\rho\,d\lambda =$$
$$= \lim_{\rho\to 0}\lim_{h\to+\infty}\int_A u\varphi_\rho\,d\lambda_h = \lim_{h\to+\infty}\int_A u\,d\lambda_h\ . \ \blacksquare$$

For every functional $L:M(\Omega;\mathbf{R}^n)\times\mathbb{A}\to[0,+\infty]$ we define as in (1.3.1)

$$SL(\lambda,A) = \inf\left\{\liminf_{h\to+\infty} L(\lambda_h,A)\ :\ \lambda_h\to\lambda\right\}\ . \tag{3.3.3}$$

LEMMA 3.3.4. *Let* $L:M(\Omega;\mathbf{R}^n)\times\mathbb{A}\to[0,+\infty]$ *be a functional, and let for every* $\varepsilon>0$

$$L_\varepsilon(\lambda,A) = L(\lambda,A) + \varepsilon\,|\lambda|(A)\,.$$

Then we have

$$SL = \inf\left\{SL_\varepsilon\ :\ \varepsilon>0\right\}\ .$$

Proof. The inequality $\le$ is trivial. To prove the opposite inequality, fix $\lambda\in M(\Omega;\mathbf{R}^n)$, $A\in\mathbb{A}$, $\delta>0$; we want to prove that

$$(3.3.4) \qquad SL(\lambda,A)+\delta \ge \inf\left\{SL_\varepsilon(\lambda,A) : \varepsilon>0\right\}.$$

By definition of $SL(\lambda,A)$ there exists a sequence $\{\lambda_h\}$ in $M(\Omega;\mathbf{R}^n)$ such that $\lambda_h\to\lambda$ and

$$SL(\lambda,A)+\delta \ge \lim_{h\to+\infty} L(\lambda_h,A).$$

Then, for every $\varepsilon>0$

$$SL(\lambda,A)+\delta \ge \liminf_{h\to+\infty}\left[L_\varepsilon(\lambda_h,A)-\varepsilon\,|\lambda_h|(A)\right] \ge$$

$$\ge SL_\varepsilon(\lambda,A)-\varepsilon\sup\left\{|\lambda_h|(A) : h\in\mathbf{N}\right\}$$

and, passing to the limit as $\varepsilon\to 0$, (3.3.4) is proved. ∎

Let Ξ be the set of all countable ordinals; for every $\xi\in\Xi$ we define by transfinite induction a functional $F_\xi:M(\Omega;\mathbf{R}^n)\times\mathbb{A}\to[0,+\infty]$ in the following way

$$(3.3.5) \qquad \begin{cases} F_0 = F \\ F_{\xi+1} = SF_\xi \\ F_\xi = \inf\{F_\eta : \eta<\xi\} \text{ if } \xi \text{ is a limit ordinal.}\end{cases}$$

By Proposition 1.3.2 we have

$$(3.3.6) \qquad \Gamma F = \inf\{F_\xi : \xi\in\Xi\} = F_\Xi .$$

Therefore, in order to prove Theorem 3.3.1 it will be enough to show that the functional F_Ξ admits an extension to $M(\Omega;\mathbf{R}^n)\times\mathbb{B}$ which satisfies all properties of the integral representation Theorem 3.2.1.

LEMMA 3.3.5. *Let* $L:M(\Omega;\mathbf{R}^n)\times\mathbb{A}\to[0,+\infty]$ *be a functional and let* $a\in L^1(\Omega)$.

Assume that

(i) L *is* $\mathbb{A}$*-local, that is*

$$A\in\mathbb{A},\ \lambda,\nu\in M(\Omega;\mathbf{R}^n),\ |\lambda-\nu|(A)=0 \quad\Rightarrow\quad L(\lambda,A)=L(\nu,A)\,;$$

(ii) *for every* $\lambda\in M(\Omega;\mathbf{R}^n)$ *the set function* $L(\lambda,\cdot)$ *is increasing and finitely additive on* $\mathbb{A}$*;*

(iii) $L(\lambda,A)=L(\lambda,B)$ *for every* $\lambda\in M(\Omega;\mathbf{R}^n)$ *and every* $A,B\in\mathbb{A}$ *with* $|\lambda|(A\Delta B)=\mu(A\Delta B)=0$;

(iv) $L(0,A)\le\int_A a\,d\mu$ *for every* $A\in\mathbb{A}$;

Then, the functional SL *defined in (3.3.3) also satisfies conditions (i),...,(iv) above. Moreover, we have*

$$\text{(3.3.7)}\qquad SL(\lambda,A) = \inf\Big\{\liminf_{h\to+\infty} L(\lambda_h,A)\ :\ \lambda_h\to\lambda \textit{ in } A\Big\}\,.$$

Proof. Property (iv) follows immediately from the definition of SL. By Lemma 3.3.4 there is no loss of generality if we assume that for a suitable constant $c>0$

$$\text{(3.3.8)}\qquad L(\lambda,A)\ \ge\ c\,|\lambda|(A) \qquad \text{for every } \lambda\in M(\Omega;\mathbf{R}^n) \text{ and } A\in\mathbb{A}.$$

We prove now formula (3.3.7). Under assumption (3.3.8), it is easy to see that the infimum in the right-hand side of (3.3.7) is actually attained. The inequality $\ge$ in (3.3.7) is trivial; in order to prove the opposite inequality, let $A\in\mathbb{A}$ and let $\{\sigma_h\}$ be a sequence in $M(\Omega;\mathbf{R}^n)$ such that $\sigma_h\to\lambda$ in A and

$$\text{(3.3.9)}\qquad \lim_{h\to+\infty} L(\sigma_h,A) = \inf\Big\{\liminf_{h\to+\infty} L(\lambda_h,A)\ :\ \lambda_h\to\lambda \text{ in } A\Big\} < +\infty\,.$$

By using (3.3.8) we may assume that $1_A\cdot|\sigma_h|\to\nu$ for a suitable $\nu\in M(\Omega)$. We claim that $\nu(\partial A)=0$. By contradiction, if $\nu(\partial A)=\gamma>0$, setting for every $\rho>0$

$$A_\rho = \{x\in\Omega\ :\ d(x,\partial A)<\rho\}$$

we have, recalling (3.1.3),

$$\nu(A_\rho) \le \liminf_{h\to+\infty} |\sigma_h|(A\cap A_\rho) .$$

The set

$$T = \left\{\rho>0 \ : \ \mu(\partial A_\rho) + \nu(\partial A_\rho) + \sum_{h=1}^{\infty} |\sigma_h|(\partial A_\rho) \neq 0\right\}$$

is at most countable, so that we may find sequences $\rho_k \to 0$ and $\sigma_{h(k)} \in M(\Omega;\mathbf{R}^n)$ such that

$$\rho_k \in]0,+\infty[-T \qquad \text{and} \qquad |\sigma_{h(k)}|(A\cap A_{\rho_k}) \ge \frac{\gamma}{2} .$$

The sequence $\nu_k = 1_{\Omega - A_{\rho_k}} \cdot \sigma_{h(k)}$ is w*-convergent to λ in A, so that

$$\liminf_{k\to+\infty} L(\nu_k,A) = \liminf_{k\to+\infty} \left[L(\sigma_{h(k)},A-\overline{A}_{\rho_k}) + L(0,A\cap A_{\rho_k})\right] =$$

$$= \liminf_{h\to+\infty} L(\sigma_{h(k)},A-\overline{A}_{\rho_k}) = \liminf_{k\to+\infty} \left[L(\sigma_{h(k)},A) - L(\sigma_{h(k)},A\cap A_{\rho_k})\right] \le$$

$$\le \liminf_{k\to+\infty} L(\sigma_{h(k)},A) - c\frac{\gamma}{2} < \lim_{h\to+\infty} L(\sigma_h,A) ,$$

which is in contradiction with (3.3.9).

Since $\nu(\partial A)=0$, by Lemma 3.3.3 we get $1_A\cdot\sigma_h \to 1_A\cdot\lambda$, so that in (3.3.7) the inequality $\le$ is also proved.

From equality (3.3.7) we immediately obtain that the functional SL is $\mathbb{A}$-local. Moreover, since for every $A,B \in \mathbb{A}$

$$\lambda_h \to \lambda \text{ in } A\cup B \ \Rightarrow\ \lambda_h \to \lambda \text{ in } A \text{ and } \lambda_h \to \lambda \text{ in } B$$

$$\lambda_h \to \lambda \text{ in } A \text{ and } \nu_h \to \nu \text{ in } B \ \Rightarrow\ 1_A\cdot\lambda_h + 1_B\cdot\nu_h \to 1_A\cdot\lambda + 1_B\cdot\nu \text{ in } A\cup B ,$$

equality (3.3.7) also implies that the set function $SL(\lambda,\cdot)$ is increasing and additive on $\mathbb{A}$ for every $\lambda \in M(\Omega;\mathbf{R}^n)$.

Finally, we prove property (iii). Let $A,B \in \mathbb{A}$ and let $\lambda \in M(\Omega;\mathbf{R}^n)$ with

$$(3.3.10) \qquad |\lambda|(A\Delta B) = \mu(A\Delta B) = 0.$$

It is not restrictive to assume $B\subset A$. By (3.3.7) and (3.3.8) there exist $\nu\in M(\Omega)$ and a sequence $\{\lambda_h\}$ with $\lambda_h\to\lambda$ in B such that

$$SL(\lambda,B) = \lim_{h\to+\infty} L(\lambda_h,B) \qquad \text{and} \qquad 1_B\cdot|\lambda_h|\to\nu .$$

Arguing as in the proof of (3.3.7) we see that $\nu(\partial B)=0$, so that Lemma 3.3.3 and (3.3.10) yield

$$1_B\cdot\lambda_h \to 1_B\cdot\lambda \qquad \text{and} \qquad 1_B\cdot\lambda_h + 1_{\Omega-A}\cdot\lambda \to \lambda .$$

Therefore

$$SL(\lambda,A) \le \lim_{h\to+\infty} L(1_B\cdot\lambda_h+1_{\Omega-A}\cdot\lambda,A) =$$

$$= \liminf_{h\to+\infty} L(1_B\cdot\lambda_h,A) = \lim_{h\to+\infty} L(\lambda_h,B) =$$

$$=SL(\lambda,B) \le SL(\lambda,A) ,$$

and this concludes the proof of the lemma. ■

In the following we shall say a set function $\gamma:\mathbb{B}\to[0,+\infty]$ is **inner regular** if

$$\gamma(B) = \sup\{\gamma(K) : K\subset B, K \text{ compact}\},$$

and **outer regular** if

$$\gamma(B) = \inf\{\gamma(A) : A\supset B, A \text{ open}\} .$$

The set function γ will be said **regular** if it is inner regular and outer regular. The following lemma will be useful (see De Giorgi & Letta [136] for further details).

LEMMA 3.3.6. *Let* $\alpha:\mathbb{A}\to[0,+\infty]$ *be an increasing set function satisfying the following conditions:*

(i) $\alpha(\emptyset) = 0$;

(ii) α *is subadditive, that is*

$$\alpha(A_1 \cup A_2) \le \alpha(A_1) + \alpha(A_2) \quad \textit{whenever } A_1, A_2 \in \mathbb{A};$$

(iii) α *is superadditive, that is*

$$\alpha(A_1 \cup A_2) \ge \alpha(A_1) + \alpha(A_2) \quad \textit{whenever } A_1, A_2 \in \mathbb{A} \textit{ with } A_1 \cap A_2 = \emptyset;$$

(iv) *there exists a finite regular measure* $\beta:\mathbb{B} \to [0,+\infty]$ *such that*

$$\alpha(A) \le \beta(A) \quad \textit{for every } A \in \mathbb{A}.$$

Then the set function $\alpha^*:\mathbb{B} \to [0,+\infty]$ *defined by*

$$\alpha^*(B) = \inf\{\alpha(A) : A \supset B,\ A \in \mathbb{A}\}$$

is a regular measure extending α *to* $\mathbb{B}$.

Proof. The fact that α^* is monotone and extends α is an immediate consequence of the monotonicity of α; moreover, the outer regularity of α^* is obvious. The rest of the proof will be given in several steps.

Step 1. *The set function* α^* *is subadditive.*

Let $B_1, B_2 \in \mathbb{B}$ and let $A_1, A_2 \in \mathbb{A}$ with $A_1 \supset B_1$ and $A_2 \supset B_2$. By the subadditivity of α we get

$$\alpha^*(B_1 \cup B_2) \le \alpha(A_1 \cup A_2) \le \alpha(A_1) + \alpha(A_2)$$

and, since A_1 and A_2 are arbitrary, we get

$$\alpha^*(B_1 \cup B_2) \le \alpha^*(B_1) + \alpha^*(B_2)\,.$$

Step 2. *The set function* α^* *is inner regular.*

Let $B \in \mathbb{B}$ and let $K \subset B$ be a compact set. By Step 1 and (iv) we have

$$\alpha^*(B) \le \alpha^*(B-K) + \alpha^*(K) \le \beta(B-K) + \alpha^*(K) \le$$

$$\le \beta(B-K) + \sup\{\alpha^*(K) : K \subset B,\ K \text{ compact}\}$$

and, since K is arbitrary and β is regular, we obtain

$$\alpha^*(B) \le \sup\{\alpha^*(K) : K \subset B,\ K \text{ compact}\}.$$

The opposite inequality being a consequence of the monotonicity of α^*, the inner regularity of α^* is proved.

Step 3. *The set function* α^* *is superadditive.*

Let $B_1, B_2 \in \mathbb{B}$ with $B_1 \cap B_2 = \emptyset$, let $K_1 \subset B_1$ and $K_2 \subset B_2$ be two compact sets, and let $A \in \mathbb{A}$ be such that $K_1 \cup K_2 \subset A$. Since two disjoint compact sets have a positive distance, it is possible to find two disjoint open sets A_1, A_2 such that $A_1 \supset K_1$, $A_2 \supset K_2$, $A \supset A_1 \cup A_2$. Then, by the superadditivity of α we have

$$\alpha^*(K_1) + \alpha^*(K_2) \leq \alpha(A_1) + \alpha(A_2) \leq \alpha(A_1 \cup A_2) \leq \alpha(A)\,.$$

Since A is arbitrary, we get

$$\alpha^*(K_1) + \alpha^*(K_2) \leq \alpha^*(K_1 \cup K_2) \leq \alpha^*(B_1 \cup B_2)\,,$$

and, since K_1 and K_2 are arbitrary, by Step 2 this implies

$$\alpha^*(B_1) + \alpha^*(B_2) \leq \alpha^*(B_1 \cup B_2)\,.$$

Step 4. *The set function* α^* *is a measure.*

It is enough to prove that α^* is continuous along increasing sequences, that is

$$\alpha^*(B_h) \uparrow \alpha^*(B) \quad \text{whenever } B_h \uparrow B.$$

Let $B_h \uparrow B$; then by Step 1 and (iv) we have

$$\alpha^*(B) \leq \alpha^*(B - B_h) + \alpha^*(B_h) \leq \beta(B - B_h) + \alpha^*(B_h) \leq$$
$$\leq \beta(B - B_h) + \sup\{\alpha^*(B_h) : h \in \mathbf{N}\},$$

so that, as $h \to +\infty$

$$\alpha^*(B) \leq \sup\{\alpha^*(B_h) \ : \ h \in \mathbf{N}\}.$$

The opposite inequality is obvious. ■

LEMMA 3.3.7. *Let* $L : M(\Omega; \mathbf{R}^n) \times \mathbb{A} \to [0, +\infty]$ *be a functional satisfying conditions (i),...,(iv) of Lemma 3.3.5 and the additional lower semicontinuity condition*

(v) $$L(\lambda,A) \le \liminf_{h\to+\infty} L(\lambda_h,A)$$

whenever $A\in\mathbb{A}$ *and* $\lambda_h\to\lambda$ *in* A. *Assume also that the functional* $L(\cdot,\Omega)$ *is convex. Then the functional* L *admits an extension* Λ *to* $M(\Omega;\mathbf{R}^n)\times\mathbb{B}$ *which satisfies hypotheses (i),...,(iv) of the integral representation Theorem 3.2.1. If in addition we have*

(vi) $L(2\lambda,A)\le 2L(\lambda,A)$ *for every* $A\in\mathbb{A}$ *and every* $\lambda\in M(\Omega;\mathbf{R}^n)$ *with* $\lambda\perp\mu$,

then Λ *satisfies all the hypotheses of Theorem 3.2.1 (with* $u_0=0$*).*

Proof. Set for every $k\in\mathbf{N}$

$$L_k(\lambda,A) = \inf\left\{L(v,A)+k\,|\lambda-v|(A)\ :\ v\in M(\Omega;\mathbf{R}^n)\right\}.$$

It is easy to see that the functionals L_k satisfy the following properties for every $\lambda,v\in M(\Omega;\mathbf{R}^n)$ and every $A,B\in\mathbb{A}$:

(3.3.11) $$0 \le L_k(\lambda,A) \le \int_A a\,d\mu + k\,|\lambda|(A)\ ;$$

(3.3.12) $$|L_k(\lambda,A)-L_k(v,A)| \le k\,|\lambda-v|(A)\ ;$$

(3.3.13) $$L_k(\lambda,A\cup B) \ge L_k(\lambda,A)+L_k(\lambda,B) \quad \text{whenever } A\cap B=\emptyset.$$

We show now that for every $\lambda\in M(\Omega;\mathbf{R}^n)$ the set function $L_k(\lambda,\cdot)$ is subadditive on $\mathbb{A}$. Fix $\lambda\in M(\Omega;\mathbf{R}^n)$ and $A,B\in\mathbb{A}$; for every $t\in\,]0,1[$ set

$$A_t = \{x\in A\cup B\ :\ t\,d(x,A-B)<(1-t)\,d(x,B-A)\}\,,$$

$$B_t = \{x\in A\cup B\ :\ t\,d(x,A-B)>(1-t)\,d(x,B-A)\}\,,$$

$$S_t = \{x\in A\cup B\ :\ t\,d(x,A-B)=(1-t)\,d(x,B-A)\}\,.$$

By the lower semicontinuity condition (v) we get that the infimum in the definition of the functional L_k is actually attained. Then, there exist two measures $v,\sigma\in M(\Omega;\mathbf{R}^n)$ such that

(3.3.14) $L_k(\lambda,A) = L(\nu,A) + k\,|\lambda-\nu|(A)$ and $L_k(\lambda,B) = L(\sigma,B) + k\,|\lambda-\sigma|(B)$.

Let $t\in]0,1[$ be such that $\mu(S_t)=|\lambda|(S_t)=|\nu|(S_t)=|\sigma|(S_t)=0$; since $A_t\subset A$ and $B_t\subset B$, by using (3.3.14) and hypothesis (iii) we obtain

$$L_k(\lambda,A\cup B) \le L(1_{A_t}\cdot\nu + 1_{B_t}\cdot\sigma,A\cup B) + k\,|\lambda-1_{A_t}\cdot\nu - 1_{B_t}\cdot\sigma|(A\cup B) \le$$

$$\le L(\nu,A_t) + L(\sigma,B_t) + k\,|\lambda-\nu|(A_t) + k\,|\lambda-\sigma|(B_t) \le$$

$$\le L(\nu,A) + L(\sigma,B) + k\,|\lambda-\nu|(A) + k\,|\lambda-\sigma|(B) = L_k(\lambda,A) + L_k(\lambda,B) .$$

We define now the functionals $\Lambda_k:M(\Omega;\mathbf{R}^n)\times\mathbb{B}\to[0,+\infty]$ by setting

$$\Lambda_k(\lambda,B) = \inf\{L_k(\lambda,A) : A\in\mathbb{A},\ A\supset B\}.$$

By (3.3.11) and Lemma 3.3.6 the set functions $\Lambda_k(\lambda,\cdot)$ are Radon measures for every $\lambda\in M(\Omega;\mathbf{R}^n)$.

We prove the functionals Λ_k are $\mathbb{B}$-local. It will be enough to show that

(3.3.15) $$\Lambda_k(\lambda,K) = \Lambda_k(\nu,K)$$

whenever K is a compact set and $|\lambda-\nu|(K)=0$. Setting for every $\rho>0$

$$K_\rho = \{x\in\Omega : d(x,K)<\rho\}$$

by (3.3.12) we have

$$|L_k(\lambda,K_\rho) - L_k(\nu,K_\rho)| \le k\,|\lambda-\nu|(K_\rho) = k\,|\lambda-\nu|(K_\rho-K) ,$$

so that

$$\Lambda_k(\lambda,K) \le L_k(\lambda,K_\rho) \le L_k(\nu,K_\rho) + k\,|\lambda-\nu|(K_\rho-K) .$$

Hence, taking the limit as $\rho\to0$,

$$\Lambda_k(\lambda,K) \le \lim_{\rho\to0} L_k(\nu,K_\rho) = \lim_{\rho\to0} \Lambda_k(\nu,K_\rho) = \Lambda_k(\nu,K) .$$

The opposite inequality can be obtained in the same way, then (3.3.15) is proved.

The functional $\Lambda = \sup\{\Lambda_k : k\in\mathbf{N}\}$ satisfies hypotheses (i),(ii),(iv) of Theorem 3.2.1; moreover, by Proposition 1.3.7, Λ is an extension of L to $M(\Omega;\mathbf{R}^n)\times\mathbb{B}$, and Λ also satisfies hypothesis (iii) of Theorem 3.2.1. In order to conclude the proof of the lemma we have to show that under assumption (vi) the functional Λ satisfies the following condition:

$$\Lambda(t\lambda,B) = t\,\Lambda(\lambda,B) + (1-t)\,\Lambda(0,B)$$

for every $t>0$, $B\in\mathbb{B}$, and $\lambda\in M(\Omega;\mathbf{R}^n)$ with $\lambda\perp\mu$. Let us fix a set $B\in\mathbb{B}$ and a measure $\lambda\in M(\Omega;\mathbf{R}^n)$ with $\lambda\perp\mu$; by the $\mathbb{B}$-locality and the additivity of Λ it is not restrictive to assume $\mu(B)=0$, therefore we have only to show that

(3.3.16) $$\Lambda(t\lambda,B) = t\,\Lambda(\lambda,B) \qquad \text{for every } t>0.$$

We claim that

(3.3.17) $$\Lambda(2t\lambda,B) \leq 2\Lambda(t\lambda,B)\,.$$

In fact, if $A\in\mathbb{A}$ with $A\supset B$, by (vi) we have

$$\Lambda(2t\lambda,B) \leq L(1_B\cdot 2t\lambda,A) \leq$$

$$\leq 2\,L(1_B\cdot t\lambda,A) = 2\,\Lambda(t\lambda,B) + 2\,\Lambda(0,A-B)$$

and, since A is arbitrary, (3.3.17) follows. Now, the function

$$\varphi(t) = \Lambda(t\lambda,B)$$

satisfies all the hypotheses of Lemma 3.1.4, therefore (3.3.16) and the final statement of the lemma are proved. ■

LEMMA 3.3.8. *Let* $L:M(\Omega;\mathbf{R}^n)\times\mathbb{A}\to[0,+\infty]$ *be a functional satisfying conditions (i),...,(iv) of Lemma 3.3.5 and the additional condition*

(3.3.18) $$L(\lambda + \nu + u\cdot\mu,A) \leq L(\lambda + u\cdot\mu,A) + L(\nu + v\cdot\mu,A)$$

for every $A\in\mathbb{A}$, $u,v\in L^1(\Omega;\mathbf{R}^n)$, *and* $\lambda,\nu\in M(\Omega;\mathbf{R}^n)$ *with* $\lambda\perp\mu$ *and* $\nu\perp\mu$. *Assume also that the functional* $L(\cdot,\Omega)$ *is convex.Then, the functional* SL *defined in (3.3.3) also satisfies condition (3.3.18).*

Proof. By Lemma 3.3.4 there is no loss of generality if we assume that there exists a constant $c>0$ such that

(3.3.19) $$L(\lambda,A) \geq c\,|\lambda|(A) \qquad \text{for every } A\in\mathbb{A} \text{ and } \lambda\in M(\Omega;\mathbf{R}^n).$$

Under assumption (3.3.19) the infimum in (3.3.7) is attained; moreover, the functional SL satisfies the following lower semicontinuity property:

(3.3.20) $$\lambda_h \to \lambda \text{ in } A \;\Rightarrow\; SL(\lambda,A) \leq \liminf_{h\to+\infty} SL(\lambda_h,A)$$

so that, by Lemmas 3.3.5 and 3.3.7, SL admits an extension Λ to $M(\Omega;\mathbf{R}^n)\times\mathbb{B}$ which satisfies hypotheses (i),...,(iv) of the integral representation Theorem 3.2.1. Let us fix $A\in\mathbb{A}$, $u,v\in L^1(\Omega;\mathbf{R}^n)$, and $\lambda,\nu\in M(\Omega;\mathbf{R}^n)$ with $\lambda\perp\mu$ and $\nu\perp\mu$; in order to prove (3.3.18) for SL it is not restrictive to assume

(3.3.21) $$SL(\lambda + u\cdot\mu,A) < +\infty \qquad \text{and} \qquad SL(\nu + v\cdot\mu,A) < +\infty\,.$$

Let M be a Borel subset of A such that

(3.3.22) $$\mu(M) = |\lambda|(A-M) = |\nu|(A-M) = 0\,;$$

then we have

$$SL(\lambda + u\cdot\mu,A) = \Lambda(\lambda,M) + \Lambda(u\cdot\mu,A-M) = \Lambda(\lambda,M) + SL(u\cdot\mu,A)\,;$$

$$SL(\nu + v\cdot\mu,A) = \Lambda(\nu,M) + \Lambda(v\cdot\mu,A-M) = \Lambda(\nu,M) + SL(v\cdot\mu,A)\,;$$

$$SL(\lambda + \nu + u\cdot\mu,A) = \Lambda(\lambda + \nu,M) + \Lambda(u\cdot\mu,A-M) =$$

$$= \Lambda(\lambda + \nu,M) + SL(u\cdot\mu,A) \leq \Lambda(\lambda + \nu,M) + SL(u\cdot\mu,A) + SL(v\cdot\mu,A)\,.$$

Therefore, the proof will be concluded if we show the following inequality

(3.3.23) $$\Lambda(\lambda + \nu,M) \leq \Lambda(\lambda,M) + \Lambda(\nu,M)\,.$$

We first prove (3.3.23) when M is a compact set. Define for every $\rho>0$

$$M_\rho = \{x\in\Omega \,:\, d(x,M)<\rho\};$$

by using (3.3.22) we get

$$\Lambda(\lambda,M) \leq SL(\lambda,M_\rho) \leq \Lambda(\lambda,M) + \Lambda(\lambda,M_\rho-M) =$$

$$= \Lambda(\lambda,M) + SL(0,M_\rho - M) \le \Lambda(\lambda,M) + \int_{M_\rho - M} a\, d\mu$$

so that

$$\Lambda(\lambda,M) = \inf\{SL(\lambda,M_\rho) : \rho>0\}.$$

Recalling (3.3.7), for every $\rho>0$ there exist a sequence $\{\lambda_h\}$ in $M(\Omega;\mathbf{R}^n)$ with $\lambda_h \to \lambda$ in M_ρ and a suitable $\alpha \in M(\Omega)$ such that

$$SL(\lambda,M_\rho) = \lim_{h\to+\infty} L(\lambda_h,M_\rho) ;$$

$$c\,|\lambda_h|(M_\rho) \le 1 + SL(\lambda,A) \qquad \text{for every } h\in\mathbf{N} ;$$

$$1_{M_\rho}\cdot|\lambda_h| \to \alpha \qquad \text{as } h\to+\infty .$$

Arguing as in the proof of Lemma 3.3.5 we get $\alpha(\partial M_\rho)=0$ for every $\rho>0$, so that Lemma 3.3.3 yields

$$1_{M_\rho}\cdot\lambda_h \to 1_{M_\rho}\cdot\lambda = \lambda \qquad \text{as } h\to+\infty .$$

The same considerations can be made for the measure ν. Consider now the set

$$T = \{\rho>0 : \mu(\partial M_\rho)>0\} ;$$

since T is at most countable, by using the facts proved above it is possible to find a sequence $\rho_h \downarrow 0$ in $]0,+\infty[-T$ and two sequences $\{\lambda_h\}$ and $\{\nu_h\}$ in $M(\Omega;\mathbf{R}^n)$ such that

(3.3.24) $$1_{M_{\rho_h}}\cdot\lambda_h \to \lambda \quad \text{and} \quad 1_{M_{\rho_h}}\cdot\nu_h \to \nu ;$$

(3.3.25) $$|\lambda_h|(\partial M_{\rho_{h+1}}) = 0 ;$$

(3.3.26) $$\left|\frac{d\lambda_h}{d\mu}\cdot\mu\right|(M_{\rho_{h+1}}) < 2^{-h} \quad \text{and} \quad \left|\frac{d\nu_h}{d\mu}\cdot\mu\right|(M_{\rho_{h+1}}) < 2^{-h} ;$$

(3.3.27) $$\Lambda(\lambda,M) = \lim_{h\to+\infty} L(\lambda_h,M_{\rho_h}) \quad \text{and} \quad \Lambda(\nu,M) = \lim_{h\to+\infty} L(\nu_h,M_{\rho_h}) .$$

For every $h \in \mathbf{N}$ let

$$\lambda_h = a_h \cdot \mu + \lambda_h^s \quad \text{and} \quad \nu_h = b_h \cdot \mu + \nu_h^s$$

be the Lebesgue-Nikodym decomposition of λ_h and ν_h (see Proposition 3.1.2), and denote for simplicity the sets M_{ρ_h} by M_h and the functions $a_h \cdot 1_{M_h - M_{h+1}}$ by α_h. By (3.3.24), (3.3.25), (3.3.26) we obtain that the sequence

$$\sigma_h = 1_{M_h} \cdot \lambda_h^s + \alpha_h \cdot \mu + 1_{M_{h+1}} \cdot \nu_{h+1}$$

tends to $\lambda + \nu$ in the weak* convergence of $M(\Omega;\mathbf{R}^n)$, so that if B is an open set with $M \subset B \subset A$, we have by the properties of L and by (3.3.18) and (3.3.25)

$$\begin{aligned} L(\sigma_h,B) &= L(\sigma_h,B-\overline{M}_h) + L(\sigma_h,M_h-\overline{M}_{h+1}) + L(\sigma_h,M_{h+1}) = \\ &= L(0,B-\overline{M}_h) + L(\lambda_h,M_h-\overline{M}_{h+1}) + L(\lambda_h^s + \nu_h^s + b_{h+1}\cdot\mu,M_{h+1}) \leq \\ &\leq L(0,B) + L(\lambda_h,M_h-\overline{M}_{h+1}) + L(\lambda_h,M_{h+1}) + L(\nu_{h+1},M_{h+1}) \leq \\ &\leq L(0,B) + L(\lambda_h,M_h) + L(\nu_{h+1},M_{h+1}) . \end{aligned}$$

Therefore, passing to the limit as $h \to +\infty$ and using (3.3.27) we get

$$SL(\lambda + \nu,B) \leq \liminf_{h\to+\infty} L(\sigma_h,B) \leq L(0,B) + \Lambda(\lambda,M) + \Lambda(\nu,M) .$$

Since B is arbitrary, (3.3.23) follows.

We prove now (3.3.23) in the general case, when M is a Borel set. Let $\{K_h\}$ be an increasing sequence of compact subsets of M such that

$$\lim_{h\to+\infty} |\lambda+\nu|(M-K_h) = 0 ;$$

then the lower semicontinuity of the functional $SL(\cdot,\Omega)$ yields

$$\begin{aligned} \Lambda(\lambda+\nu,M) + \Lambda(0,\Omega-M) &= SL(1_M\cdot(\lambda+\nu),\Omega) \leq \\ \leq \liminf_{h\to+\infty} SL(1_{K_h}\cdot(\lambda+\nu),\Omega) &= \liminf_{h\to+\infty} \left[\Lambda(\lambda+\nu,K_h) + \Lambda(0,\Omega-K_h)\right] \leq \end{aligned}$$

$$\leq \lim_{h\to+\infty}\left[\Lambda(\lambda,K_h) + \Lambda(\nu,K_h) + \Lambda(0,\Omega-M)\right] \leq \Lambda(\lambda,M) + \Lambda(\nu,M) + \Lambda(0,\Omega-M)\,.$$

Hence (3.3.23) is proved in the general case, and this concludes the proof of the lemma. ■

Proof of Theorem 3.3.1. By Remark 3.3.2 we may assume that f is a convex integrand, and up to translations we may also assume that $u_0=0$. We shall prove that the relaxed functional ΓF satisfies all the hypotheses (i),...,(vi) of Lemma 3.3.7. By (3.3.6) and Lemma 3.3.5 we get by transfinite induction that ΓF satisfies assumptions (i),...,(iv) of Lemma 3.3.7; moreover, by (3.3.7) we obtain that ΓF also satisfies assumption (v). Since the convexity condition follows from Remark 3.3.2, it remains only to prove that

$$\Gamma F(2\lambda,A) \leq 2\,\Gamma F(\lambda,A)$$

for every $A\in\mathbb{A}$ and every $\lambda\in M(\Omega;\mathbf{R}^n)$ with $\lambda\perp\mu$. But this follows, by transfinite induction, from (3.3.18) with $\nu=\lambda$ and $u=v=0$. ■

3.4. A Lower Semicontinuity Result

In this section we are concerned with the lower semicontinuity problem for functionals of the form

$$F(\lambda) = \int_\Omega f(x,\frac{d\lambda}{d\mu})\,d\mu + \int_\Omega f^\infty(x,\frac{d\lambda^s}{d|\lambda^s|})\,d|\lambda^s| \tag{3.4.1}$$

with respect to the weak* topology of $M(\Omega;\mathbf{R}^n)$. The result we shall prove is the following.

THEOREM 3.4.1. *Let* $f:\Omega\times\mathbf{R}^n\to[0,+\infty]$ *be a function such that the approximation formula*

(3.4.2) $$f(x,s) = \sup\{a_h(x) + \langle b_h(x),s\rangle : h\in\mathbf{N}\} \qquad \forall(x,s)\in\Omega\times\mathbf{R}^n$$

holds, for suitable functions $a_h\in L^1_{loc}(\Omega)$ *and* $b_h\in C(\Omega;\mathbf{R}^n)$. *Then the functional (3.4.1) is seq.* $w^*M(\Omega;\mathbf{R}^n)$*-l.s.c. on the space* $M(\Omega;\mathbf{R}^n)$.

Proof. For every $h\in\mathbf{N}$ denote for simplicity

$$f_h(x,s) = \left[a_h(x) + \langle b_h(x),s\rangle\right]^+ \qquad x\in\Omega,\ s\in\mathbf{R}^n$$

$$F_h(\lambda,B) = \int_B f_h\left(x,\frac{d\lambda}{d\mu}\right) d\mu + \int_B f_h^\infty\left(x,\frac{d\lambda^s}{d|\lambda^s|}\right) d|\lambda^s| \qquad B\in\mathbb{B},\ \lambda\in M(\Omega;\mathbf{R}^n)\ .$$

Since f is positive, we have

$$f(x,s) = \sup\{f_h(x,s) : h\in\mathbf{N}\} \qquad \forall(x,s)\in\Omega\times\mathbf{R}^n.$$

Fix $\lambda\in M(\Omega;\mathbf{R}^n)$; then by a localization argument similar to the one of Lemma 2.3.2, and by using the fact that λ^s is concentrated on a μ-negligible set, we obtain

$$F(\lambda) = \sup\left\{\sum_{i\in I} F_i(\lambda,B_i)\right\}$$

where the supremum is taken over all finite partitions of Ω into pairwise disjoint sets $B_i\in\mathbb{B}$. Since μ and $|\lambda^s|$ are regular measures we get

$$F_i(\lambda,B_i) = \sup\{F_i(\lambda,K) : K \text{ compact},\ K\subset B_i\},$$

so that

$$F(\lambda) = \sup\left\{\sum_{i\in I} F_i(\lambda,K_i) : K_i \text{ pairwise disjoint compact subsets of } \Omega\right\}.$$

By the fact that two disjoint compact sets have a positive distance, we obtain

$$F(\lambda) = \sup \Big\{ \sum_{i\in I} F_i(\lambda,A_i) \ : \ A_i \text{ pairwise disjoint open subsets of } \Omega \Big\} ,$$

so that, in order to achieve the proof, it is enough to show that for every $h\in \mathbf{N}$ and every $A\in \mathbb{A}$ the functional $F_h(\cdot,A)$ is seq. $w^*M(\Omega;\mathbf{R}^n)$-l.s.c.. By using a localization argument again, it is easy to see that

$$F_h(\lambda,A) = \sup \Big\{ \int_A a_h(x)\, \varphi(x)\, d\mu + \int_A b_h(x)\, \varphi(x)\, d\lambda \Big\}$$

where the supremum is taken over all $\varphi\in C_c(\Omega)$ with $0\le\varphi\le 1$. Therefore, the lower semicontinuity of $F_h(\cdot,A)$ follows from the continuity of the mappings

$$\lambda \to \int_A a_h(x)\, \varphi(x)\, d\mu + \int_A b_h(x)\, \varphi(x)\, d\lambda \ ,$$

and the proof is achieved. ∎

A particular case of Theorem 3.4.1 is the following.

COROLLARY 3.4.2. *Let* $f:\Omega\times\mathbf{R}^n\to[0,+\infty]$ *be a function such that*

(i) $f(x,s)$ *is lower semicontinuous in* (x,s)*;*

(ii) *for every* $x\in\Omega$ *the function* $f(x,\cdot)$ *is convex on* $\mathbf{R}^n$*;*

(iii) *one of conditions (i), (ii) of Lemma 2.2.3 (or (i') of Remark 2.2.6) is satisfied.*

Then the functional (3.4.1) is seq. $w^*M(\Omega;\mathbf{R}^n)$*-l.s.c. on the space* $M(\Omega;\mathbf{R}^n)$.

Proof. It is enough to apply Lemma 2.2.3 (or Remark 2.2.6), Lemma 2.2.5, and Theorem 3.4.1. ∎

As a corollary we obtain the following Reshetnyak lower semicontinuity theorem (see Reshetnyak [247], and also Ambrosio [11]).

THEOREM 3.4.3. *Let* $f:\Omega\times\mathbf{R}^n\to[0,+\infty]$ *be a function such that*

(a) $f(x,s)$ *is lower semicontinuous in* (x,s)*;*

(b) *for every* $x\in\Omega$ *the function* $f(x,\cdot)$ *is convex and positively 1-homogeneous.*

Then the functional

$$F(\lambda) = \int_\Omega f(x,\frac{d\lambda}{d|\lambda|})\, d|\lambda|$$

is seq. $w^*M(\Omega;\mathbf{R}^n)$*-l.s.c. on the space* $M(\Omega;\mathbf{R}^n)$.

Proof. By (a) and (b) the functional F can be written in the form

$$F(\lambda) = \int_\Omega f(x,\frac{d\lambda}{d\mu})\, d\mu + \int_\Omega f^\infty(x,\frac{d\lambda^s}{d|\lambda^s|})\, d|\lambda^s|$$

and, since $f(x,0)=0$, the result follows from Corollary 3.4.2. ■

The result of Theorem 3.4.1 may be, in a certain sense, reversed; more precisely, the following necessary condition for lower semicontinuity holds.

THEOREM 3.4.4. *Let* $f:\Omega\times\mathbf{R}^n\to[0,+\infty]$ *be a Borel function, and let* $G:L^1(\Omega;\mathbf{R}^n)\to[0,+\infty]$ *be the functional defined by*

$$G(u) = \int_\Omega f(x,u)\, d\mu\ .$$

Assume G *is sequentially l.s.c. with respect to the* $w^*M(\Omega;\mathbf{R}^n)$ *topology and not identically* $+\infty$. *Then there exist* $a_h\in L^1(\Omega)$ *and* $b_h\in C_0(\Omega;\mathbf{R}^n)$ *such that*

(3.4.3) $\quad f(x,s) = \sup\{a_h(x) + \langle b_h(x),s\rangle\ :\ h\in\mathbf{N}\}$ *for* μ*-a.e.* $x\in\Omega$ *and all* $s\in\mathbf{R}^n$.

Proof. Define for every $B\in\mathbb{B}$, $u\in L^1(\Omega;\mathbf{R}^n)$

$$G(u,B) = \int_B f(x,u)\,d\mu ,$$

and for every $k\in\mathbf{N}$

$$G_k(u,B) = \inf\left\{ G(v,B) + k\int_B |u-v|\,d\mu \ :\ v\in L^1(\Omega;\mathbf{R}^n)\right\} .$$

For every $k\in\mathbf{N}$ the functional $G_k : L^1(\Omega;\mathbf{R}^n)\times\mathbb{B}\to[0,+\infty]$ satisfies the following properties:

(i) G_k is local (in the sense of Definition 2.4.1 (a));

(ii) G_k is additive (in the sense of Definition 2.4.1 (b));

(iii) $G_k(u,B)\leq G(u,B)$ for every $u\in L^1(\Omega;\mathbf{R}^n)$ and every $B\in\mathbb{B}$;

(iv) $G_k(\cdot,\Omega)$ is seq. $w^*M(\Omega;\mathbf{R}^n)$-l.s.c. on the space $L^1(\Omega;\mathbf{R}^n)$.

Properties (i), (ii), (iii) are easy to prove. In order to prove (iv) let u_h be a sequence in $L^1(\Omega;\mathbf{R}^n)$ converging to $u\in L^1(\Omega;\mathbf{R}^n)$ in the topology $w^*M(\Omega;\mathbf{R}^n)$; we have to show that

(3.4.4) $$G_k(u,\Omega) \leq \liminf_{h\to+\infty} G_k(u_h,\Omega) .$$

We may assume that the liminf in the right-hand side of (3.4.4) is a limit and that $G_k(u_h,\Omega)$ is bounded (with respect to h). Let $v_h\in L^1(\Omega;\mathbf{R}^n)$ be such that

(3.4.5) $$G_k(u_h,\Omega) \geq G(v_h,\Omega) + k\int_\Omega |u_h-v_h|\,d\mu - \frac{1}{h} .$$

Then $\{v_h\}$ is bounded in $L^1(\Omega;\mathbf{R}^n)$ and, possibly passing to a subsequence, we may assume that $v_h\to\lambda$ in $w^*M(\Omega;\mathbf{R}^n)$ for a suitable measure $\lambda\in M(\Omega;\mathbf{R}^n)$. By the Lebesgue-Nikodym decomposition result (see Proposition 3.1.2) we have

$$\lambda = v\cdot\mu + \lambda^s$$

where $v\in L^1(\Omega;\mathbf{R}^n)$ and λ^s is singular with respect to μ. Denoting by ΓG the relaxed functional of G with respect to $w^*M(\Omega;\mathbf{R}^n)$, by Theorem 3.3.1 we have

$$\Gamma G(\lambda) = \int_\Omega \varphi(x,v)\, d\mu + \int_\Omega \varphi^\infty(x,\frac{d\lambda^s}{d|\lambda^s|})\, d|\lambda^s|$$

for a suitable convex integrand $\varphi:\Omega\times\mathbf{R}^n\to[0,+\infty]$; moreover, since G is sequentially $w^*M(\Omega;\mathbf{R}^n)$-l.s.c., it is

$$G(v) = \int_\Omega \varphi(x,v)\, d\mu .$$

Therefore, passing to the limit as $h\to+\infty$ in (3.4.5)

$$\liminf_{h\to+\infty} G_k(u_h,\Omega) \geq \Gamma G(\lambda) + k\int_\Omega |u-v|\, d\mu + k|\lambda^s|(\Omega) \geq$$

$$\geq \int_\Omega \varphi(x,v)\, d\mu + k\int_\Omega |u-v|\, d\mu =$$

$$= G(v) + k\int_\Omega |u-v|\, d\mu \geq G_k(u),$$

so that property (iv) is proved.

Then, the functional G_k satisfies all conditions of the integral representation Theorem 2.4.6 and Corollary 2.4.8, and so we have

$$(3.4.6) \qquad G_k(u,B) = \int_B f_k(x,u)\, d\mu \qquad \forall u\in L^1(\Omega;\mathbf{R}^n),\ \forall B\in\mathbb{B}$$

where $f_k(x,s)$ is a suitable convex integrand. Proposition 1.3.7 yields

$$(3.4.7) \qquad |G_k(u,B) - G_k(v,B)| \leq k\int_B |u-v|\, d\mu$$

$$(3.4.8) \qquad G_k(u,B) \uparrow G(u,B) \quad \text{as } k\to+\infty$$

so that, by (3.4.6) and Proposition 2.1.3, there exists a μ-negligible set $N\in\mathbb{B}$ such that for every $x\in\Omega-N$ and $s,t\in\mathbf{R}^n$

(3.4.9) $\qquad |f_k(x,s) - f_k(x,t)| \leq k\,|s - t|$

(3.4.10) $\qquad f_k(x,s) \uparrow f(x,s) \quad \text{as } k \to +\infty.$

Therefore, it is enough to prove (3.4.3) for each function f_k. In other words, it is not restrictive to assume on the function f that

(3.4.11) $\qquad |f(x,s) - f(x,t)| \leq c\,|s - t| \qquad \forall x \in \Omega,\ \forall s,t \in \mathbf{R}^n$

for a suitable $c>0$, and that

$$f(x,\cdot) \text{ is convex and finite for every } x \in \Omega.$$

The relaxed functional ΓG is convex and seq. $w^*M(\Omega;\mathbf{R}^n)$-l.s.c., hence it is also topologically $w^*M(\Omega;\mathbf{R}^n)$-l.s.c. by Proposition 1.1.6 (iii); therefore, by Lemma 3.1.6 for every $q \in \mathbf{Q}^n$ and $k \in \mathbf{N}$ there exists a weakly* continuous linear map $\alpha_{q,k}$ from $M(\Omega;\mathbf{R}^n)$ into $\mathbf{R}$ such that

$$\Gamma G(\lambda) \geq \Gamma G(q\cdot\mu) + \langle \alpha_{q,k}, \lambda - q\cdot\mu \rangle - {}^1/_k \qquad \forall \lambda \in M(\Omega;\mathbf{R}^n).$$

Since all weakly* continuous linear maps on $M(\Omega;\mathbf{R}^n)$ are the ones induced by the duality with $C_0(\Omega;\mathbf{R}^n)$, there exist functions $a_{q,k} \in C_0(\Omega;\mathbf{R}^n)$ such that

(3.4.12) $\qquad \Gamma G(\lambda) \geq \Gamma G(q\cdot\mu) + \langle \lambda - q\cdot\mu, a_{q,k} \rangle - {}^1/_k \qquad \forall \lambda \in M(\Omega;\mathbf{R}^n).$

Set now for every $B \in \mathbb{B}$

$$\nu_{q,k}(B) = \sup\Big\{ \Gamma G(q\cdot\mu,B) - \Gamma G(\lambda,B) + \int_B a_{q,k}\cdot d\lambda - \int_B a_{q,k}\cdot q\, d\mu \ :\ \lambda \in M(\Omega;\mathbf{R}^n) \Big\};$$

it is not difficult to check that $\nu_{q,k}$ is a positive measure, and by (3.4.12) it follows that $\nu_{q,k}(\Omega) \leq {}^1/_k$. Let

$$\nu_{q,k} = \vartheta_{q,k}\cdot\mu + \nu^s_{q,k}$$

be the Lebesgue-Nikodym decomposition of $\nu_{q,k}$ and set

$$g_{q,k}(x,s) = f(x,q) + a_{q,k}(x)\cdot(s-q) - \vartheta_{q,k}(x).$$

Since G is seq. $w^*M(\Omega;\mathbf{R}^n)$-l.s.c. on $L^1(\Omega;\mathbf{R}^n)$ we have $\Gamma G = G$ on $L^1(\Omega;\mathbf{R}^n)$, so that by the definition of $\nu_{q,k}$

$$\int_B \vartheta_{q,k}\, d\mu \ \geq \ \int_B \big[f(x,q) - f(x,u) + a_{q,k}(x)\cdot(u-q)\big]\, d\mu$$

for every $u\in L^1(\Omega;\mathbf{R}^n)$ and every $B\in\mathbb{B}$. By Proposition 2.1.3 this implies

(3.4.13) $\qquad f(x,s) \geq g_{q,k}(x,s) \qquad$ for μ-a.e. $x\in\Omega$ and for every $s\in\mathbf{R}^n$.

On the other hand, since for every $q\in\mathbf{Q}^n$ the sequence $\vartheta_{q,k}$ tends to zero strongly in $L^1(\Omega;\mathbf{R}^n)$ we have

(3.4.14) $$f(x,q) = \lim_{k\to+\infty} g_{q,k}(x,q) \leq \sup_{k\in\mathbf{N}} g_{q,k}(x,q)$$

for μ-a.e. $x\in\Omega$ and for every $q\in\mathbf{Q}^n$, so that, by (3.4.13) and (3.4.14)

(3.4.15) $$f(x,s) = \sup\{g_{q,k}(x,s) : q\in\mathbf{Q}^n,\ k\in\mathbf{N}\}$$

for μ-a.e. $x\in\Omega$ and for every $q\in\mathbf{Q}^n$. By (3.4.11) and (3.4.15) we obtain

$$|a_{q,k}(x)| \leq c \qquad \text{for } \mu\text{-a.e. } x\in\Omega \text{ and for every } q\in\mathbf{Q}^n,\ k\in\mathbf{N}$$

so that both $f(x,\cdot)$ and $\sup\{g_{q,k}(x,s) : q\in\mathbf{Q}^n, k\in\mathbf{N}\}$ are c-Lipschitz continuous for μ-a.e. $x\in\Omega$, and this, together with (3.4.15) achieves the proof. ∎

3.5. Some Examples

In this section we shall consider some interesting cases in which it is possible to find an explicit characterization of the integrand $\varphi(x,s)$ given by Theorem 3.3.1. The lower semicontinuity results of Section 3.4 will be used. As in Section 3.3, given a Borel function $f:\Omega\times\mathbf{R}^n\to[0,+\infty]$, we denote by F the functional defined on $M(\Omega;\mathbf{R}^n)\times\mathbb{B}$ by

$$F(\lambda,B) = \begin{cases} \int_B f(x,u)\, d\mu & \text{if } \lambda=u\cdot\mu \text{ with } u\in L^1(\Omega;\mathbf{R}^n) \\ +\infty & \text{otherwise,} \end{cases}$$

and by ΓF its relaxation

$$\Gamma F(\lambda,B) = \Gamma_{seq}(w^*M(\Omega;\mathbf{R}^n)^-)F(\lambda,B)\,.$$

By Theorem 3.3.1 we know that the integral representation formula

$$(3.5.1) \qquad \Gamma F(\lambda,B) = \int_B \varphi(x,\frac{d\lambda}{d\mu})\, d\mu + \int_B \varphi^\infty(x,\frac{d\lambda^s}{d|\lambda^s|})\, d|\lambda^s|$$

holds, for a suitable convex integrand φ such that

(3.5.2) $\qquad \varphi^\infty(x,s)$ is l.s.c. in (x,s).

EXAMPLE 3.5.1. *Let* $a:\Omega\to[0,+\infty]$ *be a Borel function, and let*

$$f(x,s) = a(x)\,|s| \qquad \forall(x,s)\in\Omega\times\mathbf{R}^n.$$

Then, the integrand $\varphi(x,s)$ *appearing in formula (3.5.1) is given by*

$$\varphi(x,s) = \alpha(x)\,|s|$$

where

$$(3.5.3) \qquad \alpha = \max\{\beta : \beta \textit{ is l.s.c. on } \Omega,\ \beta\le\alpha\ \mu\textit{-a.e. on } \Omega\}.$$

Proof. By Proposition 1.1.2(ii) the function α is l.s.c. on Ω and, by the Lindelöf covering theorem, α is the supremum of a countable family $\{\beta_i\}_{i\in\mathbf{N}}$ with β_i l.s.c. on Ω and $\beta_i\le a$ μ-a.e. on Ω. Then $\alpha\le a$ μ-a.e. on Ω. Setting

$$\Phi(\lambda,B) = \int_B \alpha(x)\,|\frac{d\lambda}{d\mu}|\, d\mu + \int_B \alpha(x)\, d|\lambda^s|\,,$$

by Theorem 3.4.1 we have that $\Phi(\cdot,B)$ is seq. $w^*M(\Omega;\mathbf{R}^n)$-l.s.c. and less than or equal to $F(\cdot,B)$. Therefore it is

$$\Phi(\lambda,B) \le \Gamma F(\lambda,B) \qquad \forall\lambda\in M(\Omega;\mathbf{R}^n),\ \forall B\in\mathbb{B}.$$

To conclude the proof it remains to prove the opposite inequality. It is easy to see that the integrand $\varphi(x,s)$ which gives the integral representation of ΓF in formula (3.5.1) must be of the form

$$\varphi(x,s) = \beta(x)\,|s|$$

for some function β. Since $\Gamma F\le F$, by Proposition 2.1.3 we have

$$\beta \le a \qquad \mu\text{-a.e. on } \Omega,$$

and by (3.5.2) it follows that β is l.s.c. on Ω. Therefore, by the definition of α we have $\beta\le\alpha$, so that also the inequality

$$\Phi(\lambda,B) \ge \Gamma F(\lambda,B) \qquad \forall\lambda\in M(\Omega;\mathbf{R}^n),\ \forall B\in\mathbb{B}$$

is proved. ■

REMARK 3.5.2. If $a\in L^1_{loc}(\Omega)$ formula (3.5.3) is equivalent to

$$(3.5.4) \qquad \alpha(x) = \liminf_{y\to x}\ \limsup_{\rho\to 0^+}\Big[\frac{1}{\mu(B_\rho(y))}\int_{B_\rho(y)} a\,d\mu\Big]$$

where $B_\rho(y)$ denotes the ball

$$B_\rho(y) = \{\xi\in\Omega : d(\xi,y)<\rho\}.$$

In fact, it is easy to see that formula (3.5.4) provides a l.s.c. function which is less than or equal to a μ-a.e. on Ω, and that for every l.s.c. function β with $\beta\le a$ μ-a.e. on Ω we have

$$\beta(x) \le \liminf_{y\to x}\ \limsup_{\rho\to 0^+}\Big[\frac{1}{\mu(B_\rho(y))}\int_{B_\rho(y)} a\,d\mu\Big]$$

for every $x\in\Omega$.

EXAMPLE 3.5.3. *Let* $a:\Omega\to[0,+\infty]$ *be a Borel function, let* $p>1$, *and let*

$$f(x,s) = a(x)\,|s|^p \qquad \forall (x,s)\in\Omega\times\mathbf{R}^n.$$

Then, the integrand $\varphi(x,s)$ *appearing in formula (3.5.1) is given by*

$$\varphi(x,s) = \alpha(x)\,|s|^p$$

where

$$\alpha(x) = \begin{cases} 0 & \text{if } x\notin U \\ +\infty & \text{if } x\in U \text{ and } a(x)=0 \\ a(x) & \text{if } x\in U \text{ and } a(x)\neq 0 \end{cases}$$

and U *is the greatest open subset of* Ω *such that* $a^{1/(1-p)}\in L^1_{loc}(U)$.

Proof. Set for every $(x,s)\in\Omega\times\mathbf{R}^n$, $\lambda\in M(\Omega;\mathbf{R}^n)$, $B\in\mathbb{B}$

$$\psi(x,s) = \alpha(x)\,|s|^p$$

$$\Psi(\lambda,B) = \int_B \psi(x,\frac{d\lambda}{d\mu})\,d\mu + \int_B \psi^\infty(x,\frac{d\lambda^s}{d|\lambda^s|})\,d|\lambda^s|.$$

An easy computation shows that

$$\psi(x,s) = \sup\left\{ 1_U(x)\,\xi\,|s| - 1_U(x)\,(p-1)\,\big(\alpha(x)\big)^{1/(1-p)}\big(\tfrac{\xi}{p}\big)^{p/(p-1)} : \xi\in\mathbf{Q},\ \xi\geq 0\right\}$$

and, since U is an open set and $\alpha^{1/(1-p)}\in L^1_{loc}(U)$ we may write

$$\psi(x,s) = \sup\{a_h(x) + \langle b_h(x),s\rangle : h\in\mathbf{N}\}$$

for suitable $a_h\in L^1_{loc}(\Omega)$ and $b_h\in C(\Omega;\mathbf{R}^n)$. Then Theorem 3.4.1 applies, and the functional $\Psi(\cdot,B)$ turns out to be seq. w*$M(\Omega;\mathbf{R}^n)$-l.s.c.. It is also clear that

$$\mu(U\cap\{a=0\}) = 0\,;$$

hence $\alpha\leq a$ μ-a.e. on Ω, and so

$$\Psi(\lambda,B) \leq F(\lambda,B) \qquad \forall\lambda\in M(\Omega;\mathbf{R}^n),\ \forall B\in\mathbb{B}.$$

Therefore the inequality $\Psi\leq\Gamma F$ is proved.

We prove now the opposite inequality. By the integral representation Theorem 3.3.1 formula (3.5.1) holds for a suitable convex integrand $\varphi(x,s)$ satisfying (3.5.2); moreover, since $\Gamma F \leq F$, we obtain

$$(3.5.5) \qquad 0 \leq \varphi(x,s) \leq a(x)\,|s|^p \qquad \text{for } \mu\text{-a.e. } x \in \Omega \text{ and every } s \in \mathbf{R}^n.$$

Thus, in order to prove the inequality $\Gamma F \leq \Psi$ it will be enough to show that

$$(3.5.6) \qquad \varphi(x,s) = 0 \qquad \text{for } \mu\text{-a.e. } x \in \Omega - U \text{ and every } s \in \mathbf{R}^n$$

$$(3.5.7) \qquad \varphi^\infty(x,s) = 0 \qquad \text{for every } (x,s) \in (\Omega - U) \times \mathbf{R}^n.$$

We remark that (3.5.6) follows from (3.5.7) taking into account (3.5.5) and Lemma 3.1.5. Let us prove (3.5.7). Fix $x_0 \in \Omega - U$; then we have

$$\int_{B_\rho(x_0)} a^{1/(1-p)}\, d\mu = +\infty \qquad \text{for every } \rho > 0,$$

so that

$$\lim_{\varepsilon \to 0^+} \int_{B_{\rho_\varepsilon}(x_0)} (a+\varepsilon)^{1/(1-p)}\, d\mu = +\infty$$

for a suitable sequence $\rho_\varepsilon \to 0$. Set for every $\varepsilon > 0$

$$B_\varepsilon = B_{\rho_\varepsilon}(x_0)$$

$$c_\varepsilon = \Big(\int_{B_\varepsilon} (a+\varepsilon)^{1/(1-p)}\, d\mu \Big)^{1-p}$$

$$u_\varepsilon(x) = \Big(\frac{c_\varepsilon}{a(x)+\varepsilon} \Big)^{1/(p-1)} 1_{B_\varepsilon}(x)\,.$$

We have

$$\int_\Omega u_\varepsilon\, d\mu = c_\varepsilon^{1/(p-1)} \int_{B_\varepsilon} (a+\varepsilon)^{1/(1-p)}\, d\mu = 1$$

so that for every function $v \in C_0(\Omega)$ it is

$$\inf_{B_\varepsilon} v \le \int_\Omega v\, u_\varepsilon \, d\mu \le \sup_{B_\varepsilon} v .$$

Thus $u_\varepsilon \to \delta_{x_0}$ in the topology $w^*M(\Omega)$. Moreover

$$\int_\Omega a(x)\, u_\varepsilon^p \, d\mu = c_\varepsilon^{p/(p-1)} \int_{B_\varepsilon} a(x)\big(a(x)+\varepsilon\big)^{p/(1-p)} d\mu =$$

$$= c_\varepsilon^{p/(p-1)} \int_{B_\varepsilon} \Big[\big(a(x)+\varepsilon\big)^{1/(1-p)} - \varepsilon\big(a(x)+\varepsilon\big)^{p/(1-p)}\Big] d\mu \le$$

$$\le c_\varepsilon^{p/(p-1)} \int_{B_\varepsilon} \Big[\big(a(x)+\varepsilon\big)^{1/(1-p)} + \varepsilon\big(a(x)+\varepsilon\big)^{p/(1-p)}\Big] d\mu \le$$

$$\le c_\varepsilon^{p/(p-1)} \int_{B_\varepsilon} \Big[\big(a(x)+\varepsilon\big)^{1/(1-p)} + \big(a(x)+\varepsilon\big)^{1/(1-p)}\Big] d\mu = 2c_\varepsilon .$$

Hence, for every $s \in \mathbf{R}^n$

$$\varphi^\infty(x_0,s) = \Gamma F(s\,\delta_{x_0}) \le \liminf_{\varepsilon\to 0^+} F(s\,u_\varepsilon) =$$

$$= \liminf_{\varepsilon\to 0^+} |s|^p \int_\Omega a(x)\, u_\varepsilon^p \, d\mu \le \liminf_{\varepsilon\to 0^+} 2|s|^p c_\varepsilon = 0 ,$$

and (3.5.7) is proved. ■

EXAMPLE 3.5.4. In the case $\Omega =]-1,1[$ with μ the Lebesgue measure, consider

$$f(x,s) = |x|^r\, |s|^2 \qquad (x\in\Omega,\ s\in\mathbf{R}^n).$$

Then, for the relaxed functional ΓF we have:

$$r<1 \Rightarrow \Gamma F(\lambda) = \begin{cases} \displaystyle\int_\Omega |x|^r\, |u|^2\, dx & \text{if } \lambda = u\cdot dx \text{ with } u\in L^1(\Omega;\mathbf{R}^n) \\ +\infty & \text{otherwise} \end{cases}$$

$$r\geq 1 \Rightarrow \Gamma F(\lambda) = \begin{cases} \displaystyle\int_\Omega |x|^r\,|u|^2 dx & \text{if } \lambda = u\cdot dx + c\cdot\delta_0 \text{ with } u\in L^1(\Omega;\mathbf{R}^n) \text{ and } c\in\mathbf{R}^n \\ +\infty & \text{otherwise} \end{cases}$$

where δ_0 is the Dirac measure at the point $x=0$.

EXAMPLE 3.5.5. *Let* $a:\Omega\to[0,+\infty]$ *be a function in* $L^1(\Omega)$, *and let*

$$f(x,s) = a(x)\sqrt{1+|s|^2} \qquad \textit{for every } (x,s)\in\Omega\times\mathbf{R}^n.$$

Then, the integrand $\varphi(x,s)$ *appearing in formula (3.5.1) is given by*

$$\varphi(x,s) = \begin{cases} \alpha(x)\sqrt{1+|s|^2} & \textit{if } a(x)\leq\alpha(x) \\ a(x)\sqrt{1+|s|^2} & \textit{if } a(x)>\alpha(x) \textit{ and } |s|\sqrt{a^2(x)-\alpha^2(x)}\leq\alpha(x) \\ \alpha(x)\,|s|+\sqrt{a^2(x)-\alpha^2(x)} & \textit{if } a(x)>\alpha(x) \textit{ and } |s|\sqrt{a^2(x)-\alpha^2(x)}>\alpha(x) \end{cases}$$

where α *is the function defined in (3.5.4).*

Proof. Set for every $(x,s)\in\Omega\times\mathbf{R}^n$, $\lambda\in M(\Omega;\mathbf{R}^n)$, $B\in\mathbb{B}$

$$\varphi(x,s) = \begin{cases} \alpha(x)\sqrt{1+|s|^2} & \text{if } a(x)\leq\alpha(x) \\ a(x)\sqrt{1+|s|^2} & \text{if } a(x)>\alpha(x) \text{ and } |s|\sqrt{a^2(x)-\alpha^2(x)}\leq\alpha(x) \\ \alpha(x)\,|s|+\sqrt{a^2(x)-\alpha^2(x)} & \text{if } a(x)>\alpha(x) \text{ and } |s|\sqrt{a^2(x)-\alpha^2(x)}>\alpha(x) \end{cases}$$

$$\Psi(\lambda,B) = \int_B \psi\left(x,\frac{d\lambda}{d\mu}\right) d\mu + \int_B \psi^\infty\left(x,\frac{d\lambda^s}{d|\lambda^s|}\right) d|\lambda^s|.$$

An easy computation gives

$$\psi(x,s) = \sup\left\{\xi\,\alpha(x)\,|s| + \sqrt{\left(a(x)\vee\alpha(x)\right)^2 - \xi^2\alpha^2(x)} \;:\; \xi\in\mathbf{Q}\cap[0,1]\right\},$$

and since α is l.s.c. we may write

$$\psi(x,s) = \sup\{a_h(x) + \langle b_h(x),s\rangle \;:\; h\in\mathbf{N}\}$$

for suitable $a_h \in L^1_{loc}(\Omega)$ and $b_h \in C(\Omega;\mathbf{R}^n)$. Then Theorem 3.4.1 applies, and the functional $\Psi(\cdot,B)$ turns out to be seq. $w^*M(\Omega;\mathbf{R}^n)$-l.s.c.. The inequality

$$\psi(x,s) \le a(x)\sqrt{1+|s|^2} \qquad \text{for } \mu\text{-a.e. } x\in\Omega \text{ and all } s\in\mathbf{R}^n$$

implies

$$\Psi(\lambda,B) \le F(\lambda,B) \qquad \text{for every } \lambda\in M(\Omega;\mathbf{R}^n) \text{ and } B\in\mathbb{B},$$

and so the inequality $\Psi\le\Gamma F$ is proved.

We prove now the opposite inequality. By the relaxation Theorem 3.3.1 we have

$$\Gamma F(\lambda,B) = \int_B \varphi(x,\frac{d\lambda}{d\mu})\, d\mu + \int_B \varphi^\infty(x,\frac{d\lambda^s}{d|\lambda^s|})\, d|\lambda^s|$$

and by Theorem 3.4.4 it is

(3.5.8) $$\varphi(x,s) = \sup\{c_h(x) + \langle d_h(x),s\rangle \;:\; h\in\mathbf{N}\}$$

for μ-a.e. $x\in\Omega$ and all $s\in\mathbf{R}^n$, where $c_h\in L^1(\Omega)$ and $d_h\in C_0(\Omega;\mathbf{R}^n)$ are suitable functions. Then, we have to prove that

(3.5.9) $$\varphi(x,s) \le \psi(x,s) \qquad \text{for } \mu\text{-a.e. } x\in\Omega \text{ and all } s\in\mathbf{R}^n;$$

(3.5.10) $$\varphi^\infty(x,s) \le \psi^\infty(x,s) \qquad \text{for every } (x,s)\in\Omega\times\mathbf{R}^n.$$

Let us prove (3.5.10). Since $f(x,s)\le a(x)(1+|s|)$, recalling Example 3.5.1 we have

$$\Gamma F(\lambda,B) \le \int_B a(x)\Big(1+|\frac{d\lambda}{d\mu}|\Big)\, d\mu + \int_B \alpha(x)\, d|\lambda^s|$$

and so

$$\varphi^\infty(x,s) = \Gamma F(s\,\delta_x,\{x\}) \le \alpha(x)\,|s| = \psi^\infty(x,s)\,.$$

Let us prove (3.5.9). Since

(3.5.11) $$\varphi(x,s) \le a(x)\sqrt{1+|s|^2} \qquad \text{for } \mu\text{-a.e.} x\in\Omega \text{ and all } s\in\mathbf{R}^n,$$

recalling (3.5.8) it is enough to show that for every $h\in\mathbf{N}$

(3.5.12) $$c_h(x) + \langle d_h(x),s\rangle \le \psi(x,s) \qquad \text{for } \mu\text{-a.e. } x\in\Omega \text{ and all } s\in\mathbf{R}^n.$$

Fix $h \in \mathbf{N}$; by (3.5.11) we have

$$c_h(x) + \langle d_h(x), s\rangle \le a(x)\sqrt{1+|s|^2} \qquad \text{for } \mu\text{-a.e. } x \in \Omega \text{ and all } s \in \mathbf{R}^n,$$

so that

$$(3.5.13) \qquad c_h(x) \le \sqrt{a^2(x) - |d_h(x)|^2} \qquad \text{for } \mu\text{-a.e.} x \in \Omega.$$

Moreover, by Lemma 3.1.5 and (3.5.10) it is

$$(3.5.14) \qquad |d_h(x)| \le \varphi^\infty(x, \frac{s}{|s|}) \le \alpha(x)$$

for μ-a.e. $x \in \Omega$ and all $s \in \mathbf{R}^n$, and finally (3.5.13) and (3.5.14) yield

$$c_h(x) + \langle d_h(x), s\rangle \le \sqrt{a^2(x) - |d_h(x)|^2} + |d_h(x)|\,|s| \le$$

$$\le \sup_{t \in [0,1]} \left\{ t\,\alpha(x)\,|s| + \sqrt{a^2(x) - t^2\alpha^2(x)} \right\} \le \psi(x,s)$$

which proves (3.5.12). ∎

CHAPTER 4

Functionals Defined on Sobolev Spaces

In this chapter we consider functionals of the form

$$F(u) = \int_\Omega f(x,u,Du)\,dx$$

where Ω is a bounded open subset of $\mathbf{R}^n$, the function u varies in a Sobolev space $W^{1,p}(\Omega;\mathbf{R}^m)$, and $f:\Omega\times\mathbf{R}^m\times\mathbf{R}^{mn}\to[0,+\infty]$ is an integrand. We shall investigate on the lower semicontinuity properties and on the relaxation of the functional F with respect to the weak $W^{1,p}(\Omega;\mathbf{R}^m)$ topology and the strong $L^p(\Omega;\mathbf{R}^m)$ topology. As in the previous chapters, the relaxation result will be obtained via an integral representation theorem for functionals defined on Sobolev spaces (Theorem 4.3.2).

In the following we refer to the "scalar case" to indicate the case $m=1$, and to the "vector case" to indicate the case $m>1$.

4.1. Lower Semicontinuity Results

In this section we present some classical lower semicontinuity results for functionals defined on Sobolev spaces.

Let Ω be a bounded open subset of $\mathbf{R}^n$, let $m\geq 1$ be an integer, and let $p\in[1,+\infty]$; we denote by $W^{1,p}(\Omega;\mathbf{R}^m)$ the usual Sobolev space with norm

$$\|u\|_{W^{1,p}(\Omega;\mathbf{R}^m)} = \|u\|_{L^p(\Omega;\mathbf{R}^m)} + \|Du\|_{L^p(\Omega;\mathbf{R}^{mn})}$$

and by $wW^{1,p}(\Omega;\mathbf{R}^m)$ (respectively $w^*W^{1,\infty}(\Omega;\mathbf{R}^m)$) the weak topology of $W^{1,p}(\Omega;\mathbf{R}^m)$ (respectively the weak* topology of $W^{1,\infty}(\Omega;\mathbf{R}^m)$).

Every element $z=(z_1,\dots,z_m)$ of $\mathbf{R}^{mn}$ is an m-tuple of vectors z_i of $\mathbf{R}^n$; we shall identify z with the $m\times n$ matrix $(z_{i,j})$ whose row vectors are $z_1,\dots,z_m$ $(1\le i\le m,\ 1\le j\le n)$. Vectors $x\in\mathbf{R}^n$ are regarded as column vectors, so for every $z\in\mathbf{R}^{mn}$ and every $x\in\mathbf{R}^n$ the matrix product $zx\in\mathbf{R}^m$ is the vector whose i^{th} component is $(zx)_i=\langle z_i,x\rangle$ where $\langle\cdot,\cdot\rangle$ denotes the scalar product in $\mathbf{R}^n$.

For every $u\in W^{1,p}(\Omega;\mathbf{R}^m)$ and for a.e. $x\in\Omega$ the gradient $Du(x)$ is the element of $\mathbf{R}^{mn}$ whose row vectors are $Du_1(x),\dots,Du_m(x)$; in other words $Du(x)$ is identified with the $m\times n$ matrix $(D_ju_i(x))$ $(1\le i\le m,\ 1\le j\le n)$.

From the lower semicontinuity Theorem 2.3.1 we obtain immediately the following result.

THEOREM 4.1.1. *Assume* Ω *has a Lipschitz boundary, and let* $f:\Omega\times\mathbf{R}^m\times\mathbf{R}^{mn}\to[0,+\infty]$ *be a normal-convex integrand in the sense of Definition 2.1.1(e). Then the functional*

$$(4.1.1)\qquad F(u) = \int_\Omega f(x,u,Du)\,dx$$

is sequentially $wW^{1,1}(\Omega;\mathbf{R}^m)$*-l.s.c..*

We shall see in the following that the result above is "optimal" in the scalar case, in the sense that the convexity of $f(x,s,\cdot)$ is a necessary condition for the sequential lower semicontinuity of functionals of the form (4.1.1) with respect to the weak topology of $W^{1,p}(\Omega;\mathbf{R}^m)$. In the vector case, on the contrary, it describes only a small class of all sequentially $wW^{1,p}(\Omega;\mathbf{R}^m)$-l.s.c. functionals. For this reason, following

Morrey [233], we introduce the notion of quasi-convexity.

DEFINITION 4.1.2. *A Borel function* $f:\mathbf{R}^{mn}\to[0,+\infty[$ *is said quasi-convex if*

$$f(z)\,\text{meas}(A) \;\le\; \int_A f(z + D\varphi(x))\,dx$$

for every bounded open subset A *of* $\mathbf{R}^n$*, every* $m\times n$ *matrix* z*, and every function* $\varphi\in C^1_\circ(A;\mathbf{R}^m)$.

For a systematic study concerning the properties related to quasi-convexity we refer the interested reader to the Morrey book [234], the Dacorogna books [110] and [117], and the paper by Acerbi & Fusco [5]. Here, we simply recall some of those properties which will be used in the following.

REMARK 4.1.3. It is possible to prove that when either $m=1$ or $n=1$ quasi-convexity reduces to the usual convexity. Moreover, if $f:\mathbf{R}^{mn}\to[0,+\infty[$ is quasi-convex, then for every $z\in\mathbf{R}^{mn}$ and $c\in\mathbf{R}^m$ the function $\varphi_{c,z}:\mathbf{R}^n\to[0,+\infty[$ defined by

(4.1.2) $$\varphi_{c,z}(\xi) \;=\; f(z + c\otimes\xi)$$

is convex. This implies that every quasi-convex function is locally Lipschitz on $\mathbf{R}^{mn}$. Note that if f is a quasi-convex function of class $C^2(\mathbf{R}^{mn})$, then from the convexity of the functions $\varphi_{z,c}$ defined in (4.1.2) the so-called Legendre-Hadamard condition follows:

$$\frac{\partial^2 f}{\partial z_{ij}\,\partial z_{hk}}(z_0)\;\alpha_i\alpha_h\beta_j\beta_k \;\ge\; 0$$

for all $z_0\in\mathbf{R}^{mn}$ and for all vectors $\alpha\in\mathbf{R}^m$, $\beta\in\mathbf{R}^n$.

REMARK 4.1.4. A wide class of quasi-convex functions is the class, introduced by Ball in [33], of polyconvex functions, that is the functions

$$f(z) = g(X(z)) \qquad z \in \mathbf{R}^{mn}$$

where $X(z)$ denotes the vector of all subdeterminants of the matrix z, and g is a convex function. For instance, if $n=m=2$, every polyconvex function is of the form $g(z,\det z)$ with g convex on $\mathbf{R}^4 \times \mathbf{R}$; analogously, if $n=m=3$, every polyconvex function is of the form $g(z,\operatorname{adj} z,\det z)$ with g convex on $\mathbf{R}^9 \times \mathbf{R}^9 \times \mathbf{R}$ and where by $\operatorname{adj} z$ we denote the adjugate matrix of z, that is the transpose of the matrix of co-factors of z.

The relations between quasi-convexity and lower semicontinuity are given by the following result (for the proof we refer for instance to Morrey [234], Dacorogna [110] and [117], Acerbi & Fusco [5], Marcellini [199]).

THEOREM 4.1.5. *Let* $1 \le p \le +\infty$ *and let* $f:\Omega \times \mathbf{R}^m \times \mathbf{R}^{mn} \to \mathbf{R}$ *be a Carathéodory integrand such that*

(4.1.3) $\qquad 0 \le f(x,s,z) \le a(x,|s|)\,(1 + |z|^p) \qquad$ *if* $p<+\infty$

(4.1.4) $\qquad 0 \le f(x,s,z) \le \alpha(x,|s|,|z|) \qquad$ *if* $p=+\infty$

where $a(x,t)$ *and* $\alpha(x,t,\tau)$ *are summable in* x *and increasing in* t *and* τ. *Then the following conditions are equivalent:*

(i) *for a.e.* $x \in \Omega$ *and every* $s \in \mathbf{R}^m$ *the function* $f(x,s,\cdot)$ *is quasi-convex;*

(ii) *the functional* F *given by (4.1.1) is sequentially weakly l.s.c. on* $W^{1,p}(\Omega;\mathbf{R}^m)$ *(sequentially weakly* l.s.c. on* $W^{1,\infty}(\Omega;\mathbf{R}^m)$ *if* $p=+\infty$*).*

REMARK 4.1.6. When $f(x,s,z)$ is polyconvex in z, Theorem 4.1.5 can be proved with the only assumption $0 \le f \le +\infty$ provided f is l.s.c. in z and $p \ge \min\{m,n\}$ (see Acerbi & Buttazzo & Fusco [4]). On the other hand, for general quasi-convex functions f and $p \ge 1$, it is not known if the conclusion of Theorem 4.1.5 still holds when hypotheses (4.1.3) and (4.1.4) are dropped (we refer to Dacorogna & Fusco [118] and to Marcellini [200] for a discussion on the subject).

REMARK 4.1.7. The lower semicontinuity for the integral functional

$$F(u) = \int_\Omega f(x,u,Du)\,dx \tag{4.1.5}$$

with respect to the $L^p(\Omega;\mathbf{R}^m)$ topology follows immediately from Theorems 4.1.1 or 4.1.5 provided F satisfies a growth condition of the form

$$F(u) \ge \alpha\int_\Omega |Du|^p\,dx - \beta \qquad (\alpha>0,\ \beta\ge 0,\ p>1) \tag{4.1.6}$$

for every $u \in W^{1,p}(\Omega;\mathbf{R}^m)$. But, without assumption (4.1.6), we don't know any characterization of the $L^p(\Omega;\mathbf{R}^m)$-lower semicontinuity for the functional (4.1.5) in terms of its integrand $f(x,s,z)$. Some sufficient conditions are given in the following theorem due to Serrin [262] (see also Morrey [234]).

THEOREM 4.1.8. *Let* $f:\Omega\times\mathbf{R}^m\times\mathbf{R}^{mn}\to[0,+\infty[$ *be a continuous function such that* $f(x,s,\cdot)$ *is convex for every* $(x,s)\in\Omega\times\mathbf{R}^m$. *Assume further one of the following conditions is satisfied:*

(i) $\lim_{|z|\to+\infty} f(x,s,z) = +\infty$ *for every* $(x,s)\in\Omega\times\mathbf{R}^m$;

(ii) *for every* $(x,s)\in\Omega\times\mathbf{R}^m$ *the function* $f(x,s,\cdot)$ *is stricly convex on* $\mathbf{R}^{mn}$;

(iii) *the derivatives* f_x, f_z, f_{xz} *exist and are continuous functions.*

Then, the functional F *given in (4.1.1) is l.s.c. on* $W^{1,1}_{loc}(\Omega;\mathbf{R}^m)$ *with respect to the*

$L^1_{loc}(\Omega;\mathbf{R}^m)$ *topology.*

When conditions (i), (ii), (iii) of Theorem 4.1.8 are dropped, in general the $L^1(\Omega)$-lower semicontinuity of F may fail; the following examples are known in the literature.

EXAMPLE 4.1.9. (See Carbone & Sbordone [82]). Let $n=m=1$, let $\Omega=]0,1[$, and let C be a closed subset of Ω such that $\Omega - C$ is dense in Ω and with positive measure. Consider the functional defined on $W^{1,1}(\Omega)$ by

$$F(u) = \int_\Omega f(x,u')\,dx$$

where $f(x,z) = 1_C(x)\,z^2$. By Example 3.5.3 the functional F is not $L^1(\Omega)$-l.s.c.; in fact we have for every $u \in W^{1,1}(\Omega)$

$$0 \le \Gamma(L^1(\Omega)^-)\,F(u) \le \Gamma_{seq}(wBV(\Omega)^-)\,F(u) = 0$$

where the $wBV(\Omega)$-convergence is defined in the following way:

$$v_h \to v \text{ in } wBV(\Omega) \iff v_h \to v \text{ in } L^1(\Omega) \text{ and } \{v_h\} \text{ is bounded in } W^{1,1}(\Omega)\,.$$

Nevertheless, by Theorem 4.1.1 the functional F is sequentially $wW^{1,1}(\Omega)$-l.s.c..

EXAMPLE 4.1.10. (For the proof we refer to Dal Maso [119] and Pauc [244]). Let $m=1$, $n=2$, $\Omega=]0,1[\times]0,1[$. Then it is possible to construct a continuous function $\omega:\Omega\to\mathbf{R}^2$ such that setting

$$f(x,z) = |\langle w(x),z\rangle| \qquad x\in\Omega,\ z\in\mathbf{R}^2$$

$$F(u) = \int_\Omega f(x,Du)\,dx \qquad u\in W^{1,1}(\Omega)$$

where $\langle\cdot,\cdot\rangle$ denotes the scalar product in $\mathbf{R}^2$, the functional F is not $L^1(\Omega)$-l.s.c., though by Theorem 4.1.1 it is seq. $wW^{1,1}(\Omega)$-l.s.c..

4.2. Lower Semicontinuity for Autonomous Functionals

In this section we consider autonomous functionals in the scalar case, that is functionals of the form

$$F(u) = \int_\Omega f(u,Du)\,dx \qquad u \in W^{1,1}(\Omega)$$

where $f:\mathbf{R}\times\mathbf{R}^n \to [0,+\infty[$ is a suitable function. We shall see that in this case it is possible to weaken considerably the hypotheses of Theorem 4.1.1 by requiring only the measurability of $f(s,z)$ with respect to s. More precisely, the following result has been proved by De Giorgi & Buttazzo & Dal Maso in [132].

THEOREM 4.2.1. *Let* $f:\mathbf{R}\times\mathbf{R}^n \to [0,+\infty[$ *be a function such that*

(i) *for a.e.* $s \in \mathbf{R}$ *the function* $f(s,\cdot)$ *is convex on* $\mathbf{R}^n$;

(ii) *for every* $z \in \mathbf{R}^n$ *the function* $f(\cdot,z)$ *is measurable on* $\mathbf{R}$;

(iii) *the function* $f(\cdot,0)$ *is l.s.c. on* $\mathbf{R}$;

(iv) *the function* $\alpha_f(s) = \limsup_{z\to 0} \dfrac{[f(s,0)-f(s,z)]^+}{|z|}$ *belongs to* $L^1_{loc}(\mathbf{R})$.

Then the functional

$$(4.2.1) \qquad F(u) = \int_\Omega f(u,Du)\,dx$$

is well-defined on $W^{1,1}_{loc}(\Omega)$ *and it is l.s.c. on* $W^{1,1}_{loc}(\Omega)$ *with respect to* $L^1_{loc}(\Omega)$ *topology.*

Before starting with the proof of Theorem 4.2.1 we recall some well-known properties of functions belonging to Sobolev spaces.

PROPOSITION 4.2.2. *Let* $u \in W^{1,1}_{loc}(\Omega)$ *and let* $E \subset \mathbf{R}$ *be a Borel set with* $\text{meas}(E)=0$. *Then we have* $Du=0$ *a.e. on* $u^{-1}(E)$.

Proof. See De La Vallée Poussin [137] and Serrin & Varberg [263]. ■

COROLLARY 4.2.3. *Let* $v,w \in W^{1,1}_{loc}(\Omega)$ *and let* $B \subset \Omega$ *be a Borel set such that* $v=w$ *a.e. on* B. *Then* $Dv=Dw$ *a.e. on* B.

Proof. It is enough to apply Proposition 4.2.2 with $u=v-w$ and $E=\{0\}$. ■

PROPOSITION 4.2.4. *Let* $u \in W^{1,1}_{loc}(\Omega)$ *and let* $a:\mathbf{R}\to\mathbf{R}$ *be a Lipschitz function. Then the function* $a(u)$ *belongs to* $W^{1,1}_{loc}(\Omega)$, *the function* $a'(u)Du$ *is measurable, and the usual chain-rule for derivatives holds*

$$D(a(u) = a'(u)\, Du\,.$$

Proof. See Marcus & Mizel [205], Lemmas 1.2 and 1.5. ■

REMARK 4.2.5. Note that the only hypotheses (i) and (ii) of Theorem 4.2.1 do not imply that the function $f(u,Du)$ is measurable for every $u \in W^{1,1}_{loc}(\Omega)$. Nevertheless, by using Proposition 4.2.2 and hypothesis (iii) we can prove that the functional in (4.2.1) is well-defined. In fact the following result holds.

LEMMA 4.2.6. *Let* $f:\mathbf{R}\times\mathbf{R}^n\to\mathbf{R}$ *be a function such that*

(a) *for a.e.* $s\in\mathbf{R}$ *the function* $f(s,\cdot)$ *is continuous on* $\mathbf{R}^n$;

(b) *for every* $z\in\mathbf{R}^n$ *the function* $f(\cdot,z)$ *is measurable on* $\mathbf{R}$;

(c) *the function* $f(\cdot,0)$ *is a Borel function on* $\mathbf{R}$.

Then, for every $u\in W^{1,1}_{loc}(\Omega)$ *the function* $f(u,Du)$ *is measurable on* Ω.

Proof. Let $g:\mathbf{R}\times\mathbf{R}^n\to\mathbf{R}$ be a Borel function such that

$$(4.2.2) \qquad f(s,z) = g(s,z) \qquad \text{for every } s\in\mathbf{R}-N \text{ and } z\in\mathbf{R}^n$$

where N is a suitable subset of $\mathbf{R}$ with $\text{meas}(N)=0$. Setting

$$(4.2.3) \qquad h(s,z) = \begin{cases} g(s,z) & \text{if } z\neq 0 \\ f(s,0) & \text{if } z=0\,, \end{cases}$$

hypothesis (c) implies that $h(s,z)$ is a Borel function. Let now $u\in W^{1,1}_{loc}(\Omega)$ be fixed; by Proposition 4.2.2 we have $Du=0$ a.e. on $u^{-1}(N)$, so that

$$h(u,Du) = f(u,Du) \qquad \text{a.e. on } u^{-1}(N).$$

On the other hand, by (4.2.2) and (4.2.3) we have

$$h(u(x),Du(x)) = f(u(x),Du(x)) \qquad \text{for a.e. } x\in\Omega-u^{-1}(N).$$

Then $f(u,Du)=h(u,Du)$ a.e. on Ω, and since h is a Borel function, this proves that $f(u,Du)$ is measurable. ∎

We prove now a first particular case of Theorem 4.2.1 when the function f is of the form

$$f(s,z) = [a(s) + \langle b(s),z\rangle]^+.$$

PROPOSITION 4.2.7. *Let* $a:\mathbf{R}\to\mathbf{R}$ *and* $b:\mathbf{R}\to\mathbf{R}^n$ *be two bounded measurable functions with* $a(s)\le 0$ *for every* $s\in\mathbf{R}$. *Then the functional*

$$F(u) = \int_\Omega \big[a(u) + \langle b(u),Du\rangle\big]^+ dx \tag{4.2.4}$$

is $L^1_{loc}(\Omega)$*-l.s.c. on the space* $W^{1,1}_{loc}(\Omega)$.

Proof. By Lusin's theorem there exist an increasing sequence $\{K_h\}$ of compact subsets of $\mathbf{R}$ and a sequence $\{a_h\}$ of continuous functions from $\mathbf{R}$ into $\mathbf{R}$ with $a_h\le 0$, $a_h=a$ on K_h, and $\text{meas}(\mathbf{R}-E)=0$, where $E=\bigcup\{K_h : h\in\mathbf{N}\}$. Setting

$$g(s,z) = 1_E(s)\,\big[a(s) + \langle b(s),z\rangle\big]^+$$

$$g_h(s,z) = \big[1_{K_h}(s)\, a_h(s) + \langle 1_{K_h}(s)\, b_h(s),z\rangle\big]^+$$

and using the fact that a and a_h are negative, it is easy to prove that $\{g_h\}$ is an increasing sequence of functions tending to g as $h\to+\infty$. Therefore, Beppo Levi monotone convergence theorem implies

$$\int_\Omega g(u,Du)\,dx = \sup_{h\in\mathbf{N}} \int_\Omega g_h(u,Du)\,dx \qquad \text{for every } u\in W^{1,1}_{loc}(\Omega). \tag{4.2.5}$$

Moreover, by Proposition 4.2.2, taking into account that $\text{meas}(\mathbf{R}-E)=0$, we obtain

$$F(u) = \int_\Omega g(u,Du)\,dx \qquad \text{for every } u\in W^{1,1}_{loc}(\Omega)\,. \tag{4.2.6}$$

By (4.2.5) and (4.2.6) it is enough to prove that for every $h\in\mathbf{N}$ the functional

$$F_h(u) = \int_\Omega g_h(u,Du)\,dx$$

is $L^1_{loc}(\Omega)$-l.s.c. on the space $W^{1,1}_{loc}(\Omega)$. Fix $h \in \mathbf{N}$ and set for simplicity

$$\alpha_h(s) = 1_{K_h}(s)\, a_h(s) \qquad\qquad \beta_h(s) = 1_{K_h}(s)\, b(s)\ ;$$

since $a_h \le 0$ the function α_h is l.s.c. on $\mathbf{R}$. Moreover

$$F_h(u) = \sup\Big\{ \int_\Omega \big[\alpha_h(u) + \langle \beta_h(u), Du\rangle\big]\, \varphi(x)\, dx\ :\ \varphi \in C_0^\infty(\Omega),\ 0 \le \varphi \le 1 \Big\}\ ,$$

so that we may reduce ourselves to prove that for every $\varphi \in C_\circ^\infty(\Omega)$ with $0 \le \varphi \le 1$ the functional

$$\int_\Omega \big[\alpha_h(u) + \langle \beta_h(u), Du\rangle\big]\, \varphi(x)\, dx$$

is $L^1_{loc}(\Omega)$-l.s.c. on $W^{1,1}_{loc}(\Omega)$. The lower semicontinuity of

$$\int_\Omega \alpha_h(u)\, \varphi(x)\, dx$$

follows immediately from Fatou's lemma; for the remaining part, by Proposition 4.2.4 we have

$$\int_\Omega \langle \beta_h(u), Du\rangle\, \varphi(x)\, dx = \int_\Omega \operatorname{div}\big(B_h(u)\big)\, \varphi(x)\, dx = -\int_\Omega \langle B_h(u), D\varphi(x)\rangle\, dx$$

where $B_h(s)$ is a primitive of $\beta_h(s)$. Hence the functional

$$\int_\Omega \langle \beta_h(u), Du\rangle\, \varphi(x)\, dx$$

is continuous on $W^{1,1}_{loc}(\Omega)$ with respect to the topology induced by $L^1_{loc}(\Omega)$, and this concludes the proof. ∎

By using Proposition 4.2.7 we prove now Theorem 4.2.1 under the additional condition

$$(4.2.7) \qquad f(s,0) = 0 \qquad \text{for every } s\in \mathbf{R}.$$

THEOREM 4.2.8. *Let* $f:\mathbf{R}\times\mathbf{R}^n\to[0,+\infty[$ *be a function satisfying conditions (i) and (ii) of Theorem 4.2.1, and (4.2.7). Then the functional*

$$F(u) = \int_\Omega f(u,Du)\,dx$$

is $L^1_{loc}(\Omega)$*-l.s.c. on the space* $W^{1,1}_{loc}(\Omega)$.

Proof. By Theorem 2.2.4 and Remark 2.2.5 (with $\mathbf{R}$ instead of Ω) there exist two sequences of bounded measurable functions $a_h:\mathbf{R}\to\mathbf{R}$ and $b_h:\mathbf{R}\to\mathbf{R}^n$ such that

$$(4.2.8) \qquad f(s,z) = \sup_{h\in\mathbf{N}} \left[a_h(s) + \langle b_h(s),z\rangle\right]^+ \qquad \text{for every } s\in\mathbf{R},\ z\in\mathbf{R}^n.$$

Setting for every $h\in\mathbf{N}$ $f_h(s,z) = [a_h(s) + \langle b_h(s),z\rangle]^+$ by Lemma 2.3.2 we have

$$(4.2.9) \qquad F(u) = \sup \Big\{ \sum_{i\in I} \int_{B_i} f_i(u,Du)\,dx \Big\} \qquad \text{for every } u\in W^{1,1}_{loc}(\Omega)$$

where the supremum is taken over all finite partitions of Ω by pairwise disjoint Borel sets B_i. Since

$$\int_{B_i} f_i(u,Du)\,dx = \sup \Big\{ \int_K f_i(u,Du)\,dx \ :\ K \text{ compact},\ K\subset B_i \Big\},$$

by (4.2.9) we have

$$F(u) = \sup\Big\{ \sum_{i\in I} \int_{K_i} f_i(u,Du)\,dx : K_i \text{ pairwise disjoint compact subsets of } \Omega \Big\},$$

so that, taking into account that two disjoint compact sets have a positive distance,

$$F(u) = \sup\Big\{ \sum_{i\in I} \int_{A_i} f_i(u,Du)\, dx : A_i \text{ pairwise disjoint open subsets of } \Omega \Big\}.$$

Therefore, in order to achieve the proof, it is enough to show that for every $h\in \mathbf{N}$ and every $A\in \mathbb{A}$ the functional

$$(4.2.10) \qquad \int_A f_h(u,Du)\, dx$$

is $L^1_{loc}(\Omega)$-l.s.c. on the space $W^{1,1}_{loc}(\Omega)$. By (4.2.7) and (4.2.8) we get

$$a_h(s) \leq 0 \qquad \text{for every } s\in \mathbf{R} \text{ and } h\in \mathbf{N},$$

so that the lower semicontinuity of the functionals (4.2.10) follows from Proposition 4.2.7. ■

LEMMA 4.2.9. *Let* $b\in L^1(\mathbf{R};\mathbf{R}^n)$ *and let* $a:\mathbf{R}\to\mathbf{R}^n$ *be defined by*

$$a(s) = \int_0^s b(t)\, dt \ ,$$

and let $u\in W^{1,1}_{loc}(\Omega)$ *be a function such that*

$$(4.2.11) \qquad \int_\Omega \langle b(u),Du\rangle^+\, dx < +\infty .$$

Then, for every $\varphi\in C^\infty_o(\Omega)$ *with* $\varphi\geq 0$, *the function* $\langle b(u),Du\rangle\varphi$ *belongs to* $L^1(\Omega)$ *and*

$$(4.2.12) \qquad \int_\Omega \langle b(u),Du\rangle\, \varphi\, dx = -\int_\Omega \langle a(u),D\varphi\rangle\, dx \ .$$

Proof. The fact that $\langle b(u),Du\rangle\varphi\in L^1(\Omega)$ follows from (4.2.11) and (4.2.12). Moreover, formula (4.2.12) follows from Proposition 4.2.4 if b is bounded. In the general case, for every $h \in \mathbf{N}$ set

$$b_h(s) = \begin{cases} b(s) & \text{if } |b(s)| \le h \\ 0 & \text{otherwise} \end{cases} \qquad \text{and} \qquad a_h(s) = \int_0^s b_h(t)\, dt \ .$$

Since b_h are bounded, by Proposition 4.2.4 we have

$$\int_\Omega \langle b_h(u), Du\rangle \, \varphi \, dx = -\int_\Omega \langle a_h(u), D\varphi\rangle \, dx \ ; \tag{4.2.13}$$

moreover, the continuous functions $a_h(s)$ converge pointwise to the continuous function $a(s)$, and Beppo Levi monotone convergence theorem yields

$$\int_\Omega \langle b(u), Du\rangle^- \varphi \, dx = \lim_{h\to+\infty} \int_\Omega \langle b_h(u), Du\rangle^- \varphi \, dx$$

$$\int_\Omega \langle b(u), Du\rangle^+ \varphi \, dx = \lim_{h\to+\infty} \int_\Omega \langle b_h(u), Du\rangle^+ \varphi \, dx \ .$$

Then, by (4.2.11) and (4.2.13)

$$\int_\Omega \langle b(u), Du\rangle \, \varphi \, dx = \int_\Omega \langle b(u), Du\rangle^+ \varphi \, dx - \int_\Omega \langle b(u), Du\rangle^- \varphi \, dx =$$

$$= \lim_{h\to+\infty} \int_\Omega \langle b_h(u), Du\rangle \, \varphi \, dx = \lim_{h\to+\infty} -\int_\Omega \langle a_h(u), D\varphi\rangle \, dx =$$

$$= -\int_\Omega \langle a(u), D\varphi\rangle \, dx \ . \ ■$$

LEMMA 4.2.10. *Let* $f : \mathbf{R}\times\mathbf{R}^n \to [0,+\infty[$ *be a function satisfying conditions (i) and (ii) of Theorem 4.2.1, and (4.2.7). Then, for every* $\varphi \in C_o^\infty(\Omega)$ *with* $\varphi \ge 0$ *the functional*

$$F(u) = \int_\Omega f(u, Du)\, \varphi \, dx$$

is $L^1_{loc}(\Omega)$*-l.s.c. on the space* $W^{1,1}_{loc}(\Omega)$.

Proof. For every $h,k \in \mathbf{N}$ define

$$\Omega_{h,k} = \{x \in \Omega : \varphi(x) > k\, 2^{-h}\}$$

$$\varphi_h(x) = 2^{-h} \sum_{k=1}^{4^h} 1_{\Omega_{h,k}}(x) .$$

It is easy to see that the sequence $\{\varphi_h\}$ is increasing and $\varphi = \sup\{\varphi_h : h \in \mathbf{N}\}$. Then

$$F(u) = \sup_{h \in \mathbf{N}} \int_\Omega f(u,Du)\, \varphi_h\, dx = \sup_{h \in \mathbf{N}} 2^{-h} \sum_{k=1}^{4^h} \int_{\Omega_{h,k}} f(u,Du)\, dx ,$$

so that the lower semicontinuity of F follows from Theorem 2.4.8. ■

We are now in a position to prove the general form of Theorem 4.2.1.

Proof of Theorem 4.2.1. We consider first the case $\alpha_f \in L^1(\mathbf{R})$. For every $s \in \mathbf{R}$ we denote by $\partial f(s,0)$ the subdifferential of the convex function $f(s,\cdot)$ at the point $z=0$

$$\partial f(s,0) = \{w \in \mathbf{R}^n : f(z) \geq f(0) + \langle w,z \rangle \text{ for all } z \in \mathbf{R}^n\},$$

and by $b(s)$ the element of $\partial f(s,0)$ of least norm, that is

$$|b(s)| = \min\{|w| : w \in \partial f(s,0)\}.$$

It is known (see for instance Ekeland & Temam [148], Theorem 1.2 page 236) that $b : \mathbf{R} \to \mathbf{R}^n$ is measurable and that $|b(s)| = \alpha_f(s)$ for every $s \in \mathbf{R}$. Set now for every $s \in \mathbf{R}$ and $z \in \mathbf{R}^n$

$$g(s,z) = f(s,z) - f(s,0) - \langle b(s),z \rangle,$$

so that

$$f(s,z) = g(s,z) + f(s,0) + \langle b(s),z \rangle. \tag{4.2.14}$$

Let $\{u_h\}$ be a sequence in $W^{1,1}_{loc}(\Omega)$ converging in $L^1_{loc}(\Omega)$ to a function u belonging to $W^{1,1}_{loc}(\Omega)$; we have to prove that

(4.2.15) $$F(u) \le \liminf_{h\to+\infty} F(u_h) .$$

If the right-hand side of (4.2.15) is $+\infty$ the inequality is trivial; otherwise we may assume that the liminf is a limit, and that $F(u_h)$ is bounded, that is for a suitable constant $M\ge0$

(4.2.16) $$F(u_h) \le M \qquad \text{for every } h\in \mathbf{N}.$$

Moreover, by (4.2.14) and (4.2.16) we have

$$\int_\Omega \langle b(u_h),Du_h\rangle^+ \, dx \le F(u_h) \le M \qquad \text{for every } h\in \mathbf{N} ,$$

so that, by applying Theorem 4.2.8 to the function $\langle b(s),z\rangle^+$, we get

(4.2.17) $$\int_\Omega \langle b(u),Du\rangle^+ \, dx \le \liminf_{h\to+\infty} \int_\Omega \langle b(u_h),Du_h\rangle^+ \, dx \le M .$$

Let $\varphi\in C_0^\infty(\Omega)$ with $0\le\varphi\le1$, and set for every $s\in \mathbf{R}$

$$a(s) = \int_0^s b(t)\, dt .$$

By Lemma 4.2.9 and (4.2.17)

(4.2.18) $$\int_\Omega \langle b(u),Du\rangle\, \varphi \, dx = -\int_\Omega \langle a(u),D\varphi\rangle \, dx =$$

$$= \lim_{h\to+\infty} -\int_\Omega \langle a(u_h),D\varphi\rangle \, dx = \lim_{h\to+\infty} \int_\Omega \langle b(u_h),Du_h\rangle\, \varphi \, dx .$$

By hypothesis (iii) and Fatou's lemma

(4.2.19) $$\int_\Omega f(u,0)\, \varphi \, dx \le \liminf_{h\to+\infty} \int_\Omega f(u_h,0)\, \varphi \, dx .$$

By Lemma 4.2.10, taking into account that $g(s,z)$ satisfies conditions (i) and (ii) of Theorem 4.2.1, and (4.2.7)

$$(4.2.20) \qquad \int_\Omega g(u,Du)\,\varphi\,dx \le \liminf_{h\to+\infty} \int_\Omega g(u_h,Du_h)\,\varphi\,dx .$$

From (4.2.14), (4.2.18), (4.2.19), (4.2.20) we obtain

$$\int_\Omega f(u,Du)\,\varphi\,dx \le \liminf_{h\to+\infty} \int_\Omega f(u_h,Du_h)\,\varphi\,dx ,$$

and since

$$F(u) = \sup\Big\{ \int_\Omega f(u,Du)\,\varphi\,dx \ : \ \varphi\in C_0^\infty(\Omega),\ 0\le\varphi\le 1 \Big\}$$

we get (4.2.15) and the proof is achieved in the case $\alpha_f \in L^1(\mathbf{R})$. In the general case $\alpha_f \in L^1_{loc}(\mathbf{R})$, let $\{\sigma_h\}$ be an increasing sequence of functions from $\mathbf{R}$ into $\mathbf{R}$ such that

$$\sigma_h \in C_0^\infty(\mathbf{R}) , \qquad 0\le\sigma_h\le 1 , \qquad \sup_{h\in\mathbf{N}} \sigma_h(s) = 1 \quad \text{for every } s\in\mathbf{R} .$$

For every $h\in\mathbf{N}$ define

$$f_h(s,z) = \sigma_h(s)\, f(s,z) \qquad \text{for } s\in\mathbf{R},\ z\in\mathbf{R}^n$$

$$F_h(u) = \int_\Omega f_h(u,Du)\,dx \qquad \text{for } u\in W^{1,1}_{loc}(\Omega)$$

$$\alpha_h(s) = \alpha_{f_h}(s) = \sigma_h(s)\,\alpha_f(s) \qquad \text{for } s\in\mathbf{R} .$$

Since $\alpha_h \in L^1(\mathbf{R})$ the functionals F_h are l.s.c.; moreover

$$F(u) = \sup\{F_h(u) \ : \ h\in\mathbf{N}\}$$

so that F is l.s.c. too, and the proof is complete. ∎

REMARK 4.2.11. Theorem 4.2.1 includes functionals of the form

$$\int_\Omega \Big(\sum_{i,j=1}^n a_{ij}(u)\, D_i u\, D_j u \Big)^q dx$$

where $q \geq {}^1/_2$ and $a_{ij}(s)$ are measurable functions such that

$$\sum_{i,j=1}^{n} a_{ij}(s)\, z_i\, z_j \geq 0 \qquad \text{for every } s \in \mathbf{R},\ z \in \mathbf{R}^n .$$

REMARK 4.2.12. If $f: \mathbf{R} \times \mathbf{R}^n \to [0,+\infty[$ satisfies conditions (i) and (ii) of Theorem 4.2.1, then condition (iv) is fulfilled provided there exist $\varepsilon > 0$ and $\beta \in L^1_{loc}(\mathbf{R})$ such that

$$f(s,z) \leq \beta(s) \qquad \text{for every } s \in \mathbf{R} \text{ and } z \in \mathbf{R}^n \text{ with } |z| \leq \varepsilon.$$

REMARK 4.2.13. Hypothesis (iv) in Theorem 4.2.1 cannot be dropped, as the following example shows. Let $n=1$, $\Omega =]0,1[$, and let f be defined by

$$f(s,z) = \begin{cases} \left(1 + \dfrac{z}{s}\right)^+ & \text{if } s \neq 0 \\ 1 & \text{if } s = 0 . \end{cases}$$

The function f satisfies all conditions of Theorem 4.2.1 except (iv), but the associated functional F is not $L^1_{loc}(\Omega)$-l.s.c.. In fact, for every $\varepsilon > 0$ let $u_\varepsilon(x) = \varepsilon(1-x)$; we have that $\{u_\varepsilon\}$ converges to 0 as $\varepsilon \to 0$, but $F(u_\varepsilon) = 0$ whereas $F(0) = 1$.

REMARK 4.2.14. Theorem 4.2.1 does not hold in the vector case $m > 1$ as the following example shows. Let $n=1$, $m=2$, $\Omega =]0,1[$, and let f be defined by

$$f(s,z) = \begin{cases} |z|^2 & \text{if } s_2 > 0 \\ 2|z|^2 & \text{if } s_2 \leq 0 . \end{cases} \qquad (s \in \mathbf{R}^2,\ z \in \mathbf{R}^2)$$

For every $\varepsilon > 0$ let $u_\varepsilon(x) = (x, \varepsilon x)$; we have that $\{u_\varepsilon\}$ converges to the function $u(x) = (x,0)$ as $\varepsilon \to 0$, but

$$\lim_{\varepsilon \to 0} F(u_\varepsilon) = 1 < 2 = F(u) .$$

4.3. An Integral Representation Theorem

In this section we consider functionals $F(u,A)$ defined for every function $u \in W^{1,p}(\Omega;\mathbf{R}^m)$ and every open subset A of Ω. We show that under suitable hypotheses they admit an integral representation of the form

$$F(u,A) = \int_A f(x,Du)\,dx$$

with a Carathéodory integrand $f:\Omega\times\mathbf{R}^{mn}\to\mathbf{R}$. As in the previous chapters we denote by $\mathbb{A}$ the class of all open subsets of Ω and by $\mathbb{B}$ the class of all Borel subsets of Ω. If $A_1,A_2 \in \mathbb{A}$, by $A_1 \subset\subset A_2$ we mean that $\overline{A}_1 \subset A_2$, and for every $x \in \mathbf{R}^n$ and $\rho>0$ the symbol $B_\rho(x)$ will indicate the ball

$$B_\rho(x) = \{y \in \mathbf{R}^n : |y-x|<\rho\}.$$

DEFINITION 4.3.1. *We say an integrand* $f:\Omega\times\mathbf{R}^{mn}\to\mathbf{R}$ *satisfies*

(a) *a* **growth condition of order** $p<+\infty$ *if there exist* $a \in L^1(\Omega)$ *and* $b\geq 0$ *such that*

$$|f(x,z)| \leq a(x) + b|z|^p \quad \textit{for a.e. } x\in\Omega \textit{ and for all } z\in\mathbf{R}^{mn};$$

(b) *a* **growth condition of order** ∞ *if for every* $r\geq 0$ *there exists* $a_r \in L^1(\Omega)$ *such that*

$$|f(x,z)| \leq a_r(x) \quad \textit{for a.e. } x\in\Omega \textit{ and for all } z\in\mathbf{R}^{mn} \textit{ with } |z|\leq r.$$

We say that a functional $F:W^{1,p}(\Omega;\mathbf{R}^m)\times\mathbb{A}\to\mathbf{R}$

(c) *is* **local** *on* $\mathbb{A}$ *if* $F(u,A)=F(v,A)$ *whenever* $A\in\mathbb{A}$, $u,v\in W^{1,p}(\Omega;\mathbf{R}^m)$ *and* $u=v$ *a.e. on* A;

(d) *is a* **measure** *on* $\mathbb{A}$ *if for every* $u\in W^{1,p}(\Omega;\mathbf{R}^m)$ *the set function* $F(u,\cdot)$ *is the restriction to* $\mathbb{A}$ *of a signed Borel measure;*

(e) *satisfies a* **growth condition of order** $p<+\infty$ *if there exist* $a\in L^1(\Omega)$ *and* $b\geq 0$

such that for every $A \in \mathbb{A}$ *and every* $u \in W^{1,p}(\Omega;\mathbf{R}^m)$

$$|F(u,A)| \le \int_A \left[a(x) + b|Du|^p\right] dx \,;$$

(f) *satisfies a* **growth condition of order** ∞ *if for every* $r \ge 0$ *there exists* $a_r \in L^1(\Omega)$ *such that*

$$|F(u,A)| \le \int_A a_r(x)\, dx$$

for every $A \in \mathbb{A}$ *and every* $u \in W^{1,\infty}(\Omega;\mathbf{R}^m)$ *with* $|Du| \le r$ *a.e. on* A.

We can now state the integral representation theorem.

THEOREM 4.3.2. *Let* $p \in [1,+\infty]$ *and let* $F: W^{1,p}(\Omega;\mathbf{R}^m) \times \mathbb{A} \to \mathbf{R}$ *be a functional satisfying the following properties:*

(i) F *is local on* $\mathbb{A}$;

(ii) F *is a measure on* $\mathbb{A}$;

(iii) F *satisfies a growth condition of order* p;

(iv) *for every* $A \in \mathbb{A}$, $u \in W^{1,p}(\Omega;\mathbf{R}^m)$, $c \in \mathbf{R}^m$

$$F(u+c,A) = F(u,A) \,;$$

(v) *for every* $A \in \mathbb{A}$ *the function* $F(\cdot,A)$ *is sequentially-l.s.c. with respect to the topology* $wW^{1,p}(\Omega;\mathbf{R}^m)$ *(*$w^*W^{1,\infty}(\Omega;\mathbf{R}^m)$ *if* $p=+\infty$*).*

Then there exists a Carathéodory integrand $f: \Omega \times \mathbf{R}^{mn} \to \mathbf{R}$ *such that*

(a) $f(x,z)$ *satisfies a growth condition of order* p;

(b) *for every* $A \in \mathbb{A}$ *and every* $u \in W^{1,p}(\Omega;\mathbf{R}^m)$ *the integral representation formula holds:*

$$F(u,A) = \int_A f(x,Du)\,dx .$$

REMARK 4.3.3. If $F \geq 0$, by Theorem 4.1.5 the integrand $f(x,z)$ given by Theorem 4.3.2 is a quasi-convex function with respect to z. In this case conditions (a),(b) of Theorem 4.3.2 are actually equivalent to (i),...,(v). Moreover, the integral representation result may fail if we drop hypothesis (v), as the following example shows.

EXAMPLE 4.3.4. Let $n=m=1$, $\Omega=]0,1[$, $p\in[1,+\infty]$. Let $T:W^{1,p}(\Omega)\to L^\infty(\Omega)$ be the operator

$$(Tu)(x) = \begin{cases} 1 & \text{if } \operatorname{meas}\big(\{y\in\Omega : u'(y)=u'(x)\}\big) = 0 \\ 0 & \text{otherwise} , \end{cases}$$

and let $F:W^{1,p}(\Omega)\times\mathbb{A}\to\mathbf{R}$ be the functional defined by

$$F(u,A) = \int_A (Tu)(x)\,dx .$$

By using Theorem 2.5.1 it is easy to prove that F satisfies conditions (i),...,(iv) of Theorem 4.3.2 and that for every integrand $f(x,z)$ there exist $u\in W^{1,p}(\Omega)$ and $A\in\mathbb{A}$ such that

$$F(u,A) \neq \int_A f(x,u')\,dx .$$

Proof of Theorem 4.3.2. We consider first the case $p<+\infty$. For every $z\in\mathbf{R}^{mn}$ denote by u_z the linear function $u_z(x) = zx$. By hypotheses (ii) and (iii), for every $z\in\mathbf{R}^{mn}$ the measure $F(u_z,\cdot)$ is absolutely continuous with respect to Lebesgue measure. Then, setting for every $x\in\Omega$ and $z\in\mathbf{R}^{mn}$

$$(4.3.1) \qquad f(x,z) = \limsup_{\rho\to 0^+} \frac{F(u_z, B_\rho(x))}{\operatorname{meas}(B_\rho(x))}$$

the theorems about derivation of measures guarantee that $f(x,z)$ is an integrand, that $f(\cdot,z)\in L^1(\Omega)$ for every $z\in \mathbf{R}^{mn}$, and that

$$(4.3.2) \qquad F(u_z,A) = \int_A f(x,z)\, dx$$

for every $A\in \mathbb{A}$, $z\in \mathbf{R}^{mn}$. Moreover, from hypothesis (iii) it follows that f satisfies a growth condition of order p.

We say that a function $u\in W^{1,p}(\Omega;\mathbf{R}^m)$ is piecewise affine if there exists a finite family $\{\Omega_i\}_{i\in I}$ of open subsets of Ω such that

$$u|_{\Omega_i} \text{ is affine on each } \Omega_i \qquad \text{and} \qquad \operatorname{meas}\Big(\Omega - \bigcup_{i\in I}\Omega_i\Big) = 0\,.$$

From (4.3.2) and from hypotheses (i),(ii),(iii) it follows that

$$(4.3.3) \qquad F(u,A) = \int_A f(x,Du)\, dx$$

for every piecewise affine function $u\in W^{1,p}(\Omega;\mathbf{R}^m)$ and every $A\in \mathbb{A}$. In the following lemmas we shall prove that $f(x,z)$ is a Carathéodory integrand.

LEMMA 4.3.5. *In the scalar case* $m=1$, *for every* $x\in\Omega$ *the function* $f(x,\cdot)$ *is convex on* $\mathbf{R}^n$.

Proof. Fix $x\in\Omega$, $z_1,z_2\in \mathbf{R}^n$ with $z_1\neq z_2$, $t\in\,]0,1[$, and let $z = tz_1 + (1-t)z_2$. We have to prove that

$$f(x,z) \le t\, f(x,z_1) + (1-t)\, f(x,z_2)\,.$$

By the definition of f it is enough to show that for every $\rho>0$

(4.3.4) $$F(u_z, B_\rho(x)) \le t\, F(u_{z_1}, B_\rho(x)) + (1-t)\, F(u_{z_2}, B_\rho(x)) .$$

Let us define

$$z_0 = \frac{z_2 - z_1}{|z_2 - z_1|}$$

$$\Omega^1_{hj} = \left\{ y \in \Omega : \frac{j-1}{h} < \langle y, z_0 \rangle < \frac{j-1+t}{h} \right\}$$

$$\Omega^2_{hj} = \left\{ y \in \Omega : \frac{j-1+t}{h} < \langle y, z_0 \rangle < \frac{j}{h} \right\}$$

$$\Omega^1_h = \bigcup_{j \in \mathbf{Z}} \Omega^1_{hj}$$

$$\Omega^2_h = \bigcup_{j \in \mathbf{Z}} \Omega^2_{hj}$$

where $h \in \mathbf{N}$, $j \in \mathbf{Z}$, and $\langle \cdot,\cdot \rangle$ denotes the scalar product in $\mathbf{R}^n$. Note that, as $h \to +\infty$ we have

$$1_{\Omega^1_h} \to t \ \text{ in } w^*L^\infty(\Omega) \qquad \text{and} \qquad 1_{\Omega^2_h} \to 1-t \ \text{ in } w^*L^\infty(\Omega) .$$

Let $\{u_h\}$ be the sequence of piecewise affine functions defined by

$$u_h(y) = \begin{cases} c^1_{hj} + \langle z_1, y \rangle & \text{if } y \in \Omega^1_{hj} \\ c^2_{hj} + \langle z_2, y \rangle & \text{if } y \in \Omega^2_{hj} \end{cases}$$

where

$$c^1_{hj} = \frac{(j-1)(1-t)}{h} |z_1 - z_2| \qquad \text{and} \qquad c^2_{hj} = -\frac{jt}{h} |z_1 - z_2| .$$

For every $y \in \Omega^1_{hj}$ we have

$$|u_h(y) - u_z(y)| = |c^1_{hj} + \langle z_1 - z, y \rangle| = (1-t) \left| \frac{j-1}{h} |z_1 - z_2| + \langle z_1 - z_2, y \rangle \right| =$$

$$= (1-t)\, |z_2 - z_1| \left| \frac{j-1}{h} - \langle z_0, y \rangle \right| \le \frac{t(1-t)}{h} |z_2 - z_1| .$$

Analogously, for every $y \in \Omega^2_{hj}$ we have

$$|u_h(y) - u_z(y)| \le \frac{t(1-t)}{h} |z_2 - z_1| ,$$

so that

$$\|u_h - u_z\|_{L^\infty(\Omega)} \le \frac{t(1-t)}{h} |z_2 - z_1| .$$

Since Du_h are uniformly bounded, this implies that $u_h \to u_z$ in the topology $w^*W^{1,\infty}(\Omega)$. Since u_h are piecewise affine and $F(\cdot,B_\rho(x))$ is sequentially-l.s.c. with respect to the topology $wW^{1,p}(\Omega)$, by (4.3.3) we obtain

$$F(u_z,B_\rho(x)) \le \liminf_{h\to+\infty} F(u_h,B_\rho(x)) = \liminf_{h\to+\infty} \int_{B_\rho(x)} f(y,Du_h)\,dy =$$

$$= \liminf_{h\to+\infty} \Big[\int_{B_\rho(x)\cap\Omega^1_h} f(y,z_1)\,dy + \int_{B_\rho(x)\cap\Omega^2_h} f(y,z_2)\,dy \Big] =$$

$$= t \int_{B_\rho(x)} f(y,z_1)\,dy + (1-t) \int_{B_\rho(x)} f(y,z_2)\,dy = t\,F(u_{z_1},B_\rho(x)) + (1-t)\,F(u_{z_2},B_\rho(x))$$

which proves (4.3.4). ∎

LEMMA 4.3.6. *In the general case* $m>1$, *for a.e.* $x \in \Omega$ *the function* $f(x,\cdot)$ *is locally Lipschitz on* $\mathbf{R}^{mn}$.

Proof. By Lemma 4.3.5 for every $i=1,\ldots,m$ the function

(4.3.5) $\qquad z_i \to f(x,z_1,\ldots,z_{i-1},z_i,z_{i+1},\ldots,z_m)$

is convex on $\mathbf{R}^n$ whenever $x \in \Omega$ and $z_j \in \mathbf{R}^n$ (with $1 \le j \le m$ and $j \ne i$) are fixed. Since f satisfies a growth condition of order p, the function in (4.3.5) is locally Lipschitz on $\mathbf{R}^n$ for a.e. $x \in \Omega$, and since i is arbitrary, this implies that also $f(x,\cdot)$

is locally Lipschitz on $\mathbf{R}^{mn}$ for a.e. $x \in \Omega$. ∎

Proof of Theorem 4.3.2 (continuation). We have proved that the function $f(x,z)$ defined in (4.3.1) is a Carathéodory integrand which satisfies a growth condition of order p; therefore (see for instance Vainberg [282], Theorem 19.1) the functional

$$u \to \int_A f(x,Du)\, dx$$

is strongly continuous on $W^{1,p}(\Omega;\mathbf{R}^m)$ for every $A \in \mathbb{A}$. Let $u \in W^{1,p}(\Omega;\mathbf{R}^m)$, let $A \in \mathbb{A}$ with $A \subset\subset \Omega$, and let $\hat{u} \in W^{1,p}(\Omega;\mathbf{R}^m)$ be with compact support in Ω and such that $u=\hat{u}$ on A. We may find a sequence $\{u_h\}$ of piecewise affine functions of $W^{1,p}(\Omega;\mathbf{R}^m)$ which converges to $\hat{u}$ strongly in $W^{1,p}(\Omega;\mathbf{R}^m)$; since $F(\cdot,A)$ is seq. $wW^{1,p}(\Omega;\mathbf{R}^m)$-l.s.c., by using (4.3.3) we obtain

$$\begin{aligned} F(u,A) &= F(\hat{u},A) \leq \liminf_{h\to+\infty} F(u_h,A) = \\ &= \lim_{h\to+\infty} \int_A f(x,Du_h)\, dx = \int_A f(x,D\hat{u})\, dx = \int_A f(x,Du)\, dx \ , \end{aligned}$$

so that

$$F(u,A) \leq \int_A f(x,Du)\, dx \tag{4.3.6}$$

for every $u \in W^{1,p}(\Omega;\mathbf{R}^m)$ and every $A \in \mathbb{A}$ with $A \subset\subset \Omega$. In order to prove the opposite inequality, let us fix $u \in W^{1,p}(\Omega;\mathbf{R}^m)$ and denote by $G: W^{1,p}(\Omega;\mathbf{R}^m) \times \mathbb{A} \to \mathbf{R}$ the functional

$$G(v,A) = F(v+u,A) \,.$$

The functional G satisfies conditions (i),...,(v) of Theorem 4.3.2; therefore from (4.3.3) and (4.3.6) it follows there exists a Carathéodory integrand $g(x,z)$ satisfying a growth condition of order p, such that

(4.3.7) $$G(v,A) = \int_A g(x,Dv)\,dx$$

for every $A \in \mathbb{A}$ and every piecewise affine function $v \in W^{1,p}(\Omega;\mathbf{R}^m)$, and

(4.3.8) $$G(v,A) \leq \int_A g(x,Dv)\,dx$$

for every $A \in \mathbb{A}$ with $A \subset\subset \Omega$ and every $v \in W^{1,p}(\Omega;\mathbf{R}^m)$. Let now $A \in \mathbb{A}$ be fixed with $A \subset\subset \Omega$, and let $\hat{u} \in W^{1,p}(\Omega;\mathbf{R}^m)$ be with compact support in Ω and such that $\hat{u}=u$ on A. We may find a sequence $\{u_h\}$ of piecewise affine functions belonging to $W^{1,p}(\Omega;\mathbf{R}^m)$ which converges to $\hat{u}$ strongly in $W^{1,p}(\Omega;\mathbf{R}^m)$; then, by using (4.3.3), (4.3.6), (4.3.7), (4.3.8) we obtain

$$\int_A g(x,0)\,dx = G(0,A) = F(u,A) \leq \int_A f(x,Du)\,dx =$$

$$= \int_A f(x,D\hat{u})\,dx = \lim_{h\to+\infty} \int_A f(x,Du_h)\,dx = \lim_{h\to+\infty} F(u_h,A) =$$

$$= \lim_{h\to+\infty} G(u_h-u,A) \leq \lim_{h\to+\infty} \int_A g(x,Du_h-Du)\,dx =$$

$$= \lim_{h\to+\infty} \int_A g(x,Du_h-D\hat{u})\,dx = \int_A g(x,0)\,dx .$$

Hence

(4.3.9) $$F(u,A) = \int_A f(x,Du)\,dx$$

for every $u \in W^{1,p}(\Omega;\mathbf{R}^m)$ and every $A \in \mathbb{A}$ with $A \subset\subset \Omega$. By property (ii) it follows that (4.3.9) holds for every $u \in W^{1,p}(\Omega;\mathbf{R}^m)$ and every $A \in \mathbb{A}$, and the theorem is proved in the case $p<+\infty$.

In the case $p=+\infty$ the only change in the proof is in the choice of the sequences $\{u_h\}$ used to prove (4.3.6) and (4.3.9). In this case we use the fact that for every function

$u \in W^{1,\infty}(\Omega;\mathbf{R}^m)$ with compact support in Ω there exists a sequence $\{u_h\}$ of piecewise affine functions of $W^{1,\infty}(\Omega;\mathbf{R}^m)$ such that (see for instance Ekeland & Temam [148], Proposition 2.9, page 317)

$$\begin{cases} u_h \to u & \text{in } L^\infty(\Omega;\mathbf{R}^m) \\ Du_h(x) \to Du(x) & \text{a.e. in } \Omega \\ \|Du_h\|_{L^\infty(\Omega)} \le \|Du\|_{L^\infty(\Omega)} & \text{for every } h \in \mathbf{N}\,. \end{cases}$$

This allow us to prove formulas similar to (4.3.6) and (4.3.9) also in the case $p=+\infty$, and so the proof is concluded. ■

4.4. Relaxation in Sobolev Spaces

In this section we study the relaxation for functionals of the form

$$F(u) = \int_\Omega f(x,Du)\,dx \tag{4.4.1}$$

where $f:\Omega\times\mathbf{R}^{mn}\to[0,+\infty[$ is an integrand. Problems of this kind have been considered by many authors in a more or less general framework (see Bibliography); we shall try to present here an unified version of all the known results in the case of functionals of the form (4.4.1). To do this, we define for every $u \in W^{1,p}(\Omega;\mathbf{R}^m)$ and every $A \in \mathbb{A}$

$$F(u,A) = \int_A f(x,Du)\,dx \tag{4.4.2}$$

$$\Gamma F(u,A) = \Gamma_{seq}(wW^{1,p}(\Omega;\mathbf{R}^m)^-)\,F(u,A) \quad \left(w^*W^{1,\infty}(\Omega;\mathbf{R}^m) \text{ if } p=+\infty\right). \tag{4.4.3}$$

We shall prove the following theorem.

THEOREM 4.4.1. *Assume* $f:\Omega\times\mathbf{R}^{mn}\to[0,+\infty[$ *is an integrand satisfying a growth condition of order* p; *then there exists a positive Carathéodory integrand* $\varphi(x,z)$ *quasi-convex in* z *such that*

(i) $\varphi(x,z)$ *satisfies a growth condition of order* p;

(ii) $\Gamma F(u,A) = \int_A \varphi(x,Du)\,dx$ *for every* $u\in W^{1,p}(\Omega;\mathbf{R}^m)$ *and every* $A\in\mathbb{A}$.

We shall prove in the following lemmas that the functional ΓF satisfies all conditions of the integral representation Theorem 4.3.2.

LEMMA 4.4.2. *Let* $G:W^{1,p}(\Omega;\mathbf{R}^m)\times\mathbb{A}\to[0,+\infty[$ *be a functional satisfying conditions (i), (ii), (iii) of the integral representation Theorem 4.3.2. Let us define a functional* $SG:W^{1,p}(\Omega;\mathbf{R}^m)\times\mathbb{A}\to[0,+\infty[$ *by*

(4.4.4) $$SG(u,A) = \inf\left\{\liminf_{h\to+\infty} G(u_h,A)\right\}$$

where the infimum is taken over all sequences $\{u_h\}$ *converging to* u *in the topology* $wW^{1,p}(\Omega;\mathbf{R}^m)$ *(* $w^*W^{1,\infty}(\Omega;\mathbf{R}^m)$ *if* $p=+\infty$*). Then the functional* SG *also satisfies conditions (i), (ii), (iii) of Theorem 4.3.2.*

Proof. We prove the lemma in the case $p<+\infty$; the case $p=+\infty$ can be proved with the modifications considered in the proof of Theorem 4.3.2. Property (iii) for SG follows immediately by taking $u_h=u$ in (4.4.4). Let us prove that

(4.4.5) $$SG(u,A'\cup B) \le SG(u,A'') + SG(u,B)$$

for every $u\in W^{1,p}(\Omega;\mathbf{R}^m)$ and every $A',A'',B\in\mathbb{A}$ with $A'\subset\subset A''$. Let us fix $u\in W^{1,p}(\Omega;\mathbf{R}^m)$ and $A',A'',B\in\mathbb{A}$ with $A'\subset\subset A''$; for every $\varepsilon>0$ there exist two sequences $\{u_h\}$ and $\{v_h\}$ converging to u in $wW^{1,p}(\Omega;\mathbf{R}^m)$ such that

$$
(4.4.6)\qquad \begin{cases} SG(u,A'')+\varepsilon > \lim\limits_{h\to+\infty} G(u_h,A'') \\ SG(u,B)+\varepsilon > \lim\limits_{h\to\infty} G(v_h,B) \ . \end{cases}
$$

For every $k\in\mathbf{N}$ let $A_0, A_1,\dots,A_k$ be open sets, with $\text{meas}(\partial A_i)=0$, such that

$$
A' = A_0\subset\subset A_1\subset\subset\dots\subset\subset A_k\subset\subset A'' ,
$$

and set $S_i = A_i - \overline{A}_{i-1}$ for $i=1,\dots,k$, and $S = A_k - \overline{A}_0$. For every $i=1,\dots,k$ there exists $\sigma_i\in C_o^\infty(A_i)$ such that $0\le\sigma_i\le1$ and $\sigma_i=1$ on A_{i-1}. We define

$$
N_k = \sup\{|D\sigma_i(x)|^p : x\in\Omega, i=1,\dots,k\} \qquad (k\in\mathbf{N})
$$

$$
w_{h,i} = \sigma_i u_h + (1-\sigma_i)v_h \qquad (h\in\mathbf{N},\ 1\le i\le k).
$$

The functions $w_{h,i}$ are in $W^{1,p}(\Omega;\mathbf{R}^m)$, and for every $i\in\{1,\dots,k\}$ the sequence $\{w_{h,i}\}$ converges to u in $wW^{1,p}(\Omega;\mathbf{R}^m)$. Therefore we have

$$
(4.4.7)\qquad G(w_{h,i},A'\cup B) \le G(w_{h,i},(A'\cup B)\cap A_{i-1}) + G(w_{h,i},B-\overline{A}_i) +
$$

$$
+ G(w_{h,i},B\cap S_i) \le G(u_h,A'') + G(v_h,B) + \int_{S_i}\left[a(x) + b|Dw_{h,i}|^p\right] dx \le
$$

$$
\le G(u_h,A'')+G(v_h,B)+\int_{S_i}\left[a(x)+2^{p-1}b\left(|Du_h|^p+|Dv_h|^p+N_k|u_h-v_h|^p\right)\right]dx
$$

for suitable $a\in L^1(\Omega)$ and $b\in\mathbf{R}$ (given by Definition 4.3.1(e)). Since $\{u_h\}$ and $\{v_h\}$ converge to u in $wW^{1,p}(\Omega;\mathbf{R}^m)$ we have for a suitable constant $M>0$

$$
\int_\Omega\left[|Du_h|^p + |Dv_h|^p\right] dx \le M .
$$

Choose $k\in\mathbf{N}$ such that

$$(4.4.8) \qquad 2^{p-1}bM + \int_\Omega a(x)\,dx < k\varepsilon$$

and for every $h\in\mathbf{N}$ let $i(h)\in\{1,\dots,k\}$ be an index such that

$$\int_{S_{i(h)}} \left[a(x) + 2^{p-1}b\left(|Du_h|^p + |Dv_h|^p + N_k|u_h-v_h|^p\right)\right] dx \le$$

$$\le \frac{1}{k} \int_S \left[a(x) + 2^{p-1}b\left(|Du_h|^p + |Dv_h|^p + N_k|u_h-v_h|^p\right)\right] dx .$$

Then, by (4.4.6) we have

$$G(w_{h,i(h)},A'\cup B) \le G(u_h,A'') + G(v_h,B) + \frac{1}{k}\int_\Omega a(x)\,dx +$$

$$+ \frac{2^{p-1}bM}{k} + \frac{2^{p-1}bN_k}{k}\int_S |u_h-v_h|^p\,dx .$$

Since $\{w_{h,i(h)}\}$ converges to u in $wW^{1,p}(\Omega;\mathbf{R}^m)$ and $\{u_h-v_h\}$ converges to 0 in $L^p(S;\mathbf{R}^m)$, taking the limit as $h\to+\infty$, and using (4.4.6), (4.4.7), (4.4.8) we obtain

$$SG(u,A'\cup B) \le \liminf_{h\to+\infty} G(w_{h,i(h)},A'\cup B) \le SG(u,A'') + SG(u,B) + 3\varepsilon ,$$

and, since ε was arbitrary, this yields (4.4.5).

Let us now prove property (ii) of Theorem 4.3.2 for the functional SG. Let us fix $u\in W^{1,p}(\Omega;\mathbf{R}^m)$; then, by Lemma 3.3.6 it is enough to prove that the set function $SG(u,\cdot)$ satisfies the following properties whenever $A,A_1,A_2\in \mathbb{A}$:

(4.4.9) $\quad SG(u,A_1\cup A_2) \ge SG(u,A_1) + SG(u,A_2) \quad$ whenever $A_1\cap A_2=\emptyset$;

(4.4.10) $\quad SG(u,A_1\cup A_2) \le SG(u,A_1) + SG(u,A_2)$;

(4.4.11) $\quad 0 \le SG(u,A) \le \int_A \left[a(x) + b|Du|^p\right] dx$.

Properties (4.4.9) and (4.4.11) follow immediately from the definition of the functional SG and from the analogous properties of G. Let us prove property (4.4.10). Fix $A_1,A_2 \in \mathbb{A}$, let K be a compact set with $K \subset A_1 \cup A_2$, and let U,V be open sets with $K \subset V \subset\subset U \subset\subset A_1 \cup A_2$. We apply (4.4.5) with $A'=V$, $A''=U$, $B=(A_1 \cup A_2)-K$ and we get

(4.4.12) $\qquad SG(u,A_1 \cup A_2) \le SG(u,U) + SG(u,(A_1 \cup A_2)-K)$.

Moreover, we can find an open set $C \subset\subset A_1$ such that $U \subset C \cup A_2$, and applying (4.4.5) with $A'=C$, $A''=A_1$, $B=A_2$ we obtain

$$SG(u,U) \le SG(u,C \cup A_2) \le SG(u,A_1) + SG(u,A_2) ,$$

which, together with (4.4.12) and (4.4.11) gives

$$SG(u,A_1 \cup A_2) \le SG(u,A_1) + SG(u,A_2) + \int_{(A_1 \cup A_2)-K} \left[a(x) + b|Du|^p\right] dx .$$

Since K was arbitrary, this yields (4.4.10), so that property (ii) for SG is proved.

Let us prove the locality property (i). Fix $A \in \mathbb{A}$ and $u,v \in W^{1,p}(\Omega;\mathbf{R}^m)$ with u=v a.e. on A. For every sequence $\{u_h\}$ converging to u in $wW^{1,p}(\Omega;\mathbf{R}^m)$ and every open set $A' \subset\subset A$ we may easily construct a sequence $\{v_h\}$ converging to v in the topology $wW^{1,p}(\Omega;\mathbf{R}^m)$ such that $u_h=v_h$ on A'. Then

$$\liminf_{h\to+\infty} G(u_h,A) \ge \liminf_{h\to+\infty} G(u_h,A') =$$

$$= \liminf_{h\to+\infty} G(v_h,A') \ge SG(v,A) ,$$

and, since $\{u_h\}$ was arbitrary,

$$SG(u,A) \ge SG(v,A') .$$

But we have just proved that the functional SG is a measure on $\mathbb{A}$, so that, taking the supremum as $A' \subset\subset A$, we get

$$SG(u,A) \ge SG(v,A) .$$

The opposite inequality can be proved in the same way. ■

LEMMA 4.4.3. *The functional* ΓF *defined in (4.4.3) satisfies properties (i), (ii), (iii) of Theorem 4.3.2.*

Proof. Let Λ be the set of all countable ordinals. For every $\lambda \in \Lambda$ we define the functional $F_\lambda : W^{1,p}(\Omega;\mathbf{R}^m) \times \mathbb{A} \to \mathbf{R}$ by transfinite induction:

$$\begin{cases} F_0(u,A) = F(u,A) \\ F_{\lambda+1}(u,A) = SF_\lambda(u,A) \\ F_\lambda(u,A) = \inf\{F_\mu(u,A) : \mu<\lambda\} \quad \text{if } \lambda \text{ is a limit ordinal.} \end{cases}$$

By Proposition 1.3.2 it is

$$\Gamma F(u,A) = \inf\{F_\lambda(u,A) : \lambda \in \Lambda\}. \tag{4.4.13}$$

Let us prove, by transfinite induction, that F_λ satisfies conditions (i), (ii), (iii) of Theorem 4.3.2 for every $\lambda \in \Lambda - \{0\}$. This is true for $F_1 = SF$ by Lemma 4.4.2; if it is true for F_λ, then it is also true for $F_{\lambda+1} = SF_\lambda$ by Lemma 4.4.2 again; finally, if λ is a limit ordinal and F_μ satisfies conditions (i), (ii), (iii) for every $\mu<\lambda$, then F_λ also satisfies conditions (i), (ii), (iii): conditions (i) and (ii) are trivial, condition (iii) follows from the fact that the pointwise infimum of a decreasing family of finite non-negative measures is still a nonnegative measure.

So F_λ satisfies conditions (i), (ii), (iii) for every $\lambda \in \Lambda - \{0\}$. From (4.4.13) it follows that the same properties hold for ΓF. ∎

Proof of Theorem 4.4.1. We have to prove that the functional ΓF satisfies conditions (i),...,(v) of the integral representation Theorem 4.3.2. Conditions (i), (ii), (iii) follow from Lemma 4.4.3, and condition (v) follows immediately from the definition of ΓF. In order to prove condition (iv), let us fix $A \in \mathbb{A}$ and $c \in \mathbf{R}^m$; since $\Gamma F \le F$ we have

$$\Gamma F(u+c,A) \le F(u+c,A) = F(u,A) \qquad \text{for every } u \in W^{1,p}(\Omega;\mathbf{R}^m),$$

so that, since ΓF is seq. $wW^{1,p}(\Omega;\mathbf{R}^m)$-l.s.c.,

$$\Gamma F(u+c,A) \le \Gamma F(u,A) \qquad \text{for every } u\in W^{1,p}(\Omega;\mathbf{R}^m) \text{ and } c\in\mathbf{R}^m.$$

Taking $u=v+d$ and $c=-d$ we get

$$\Gamma F(v,A) \le \Gamma F(v+d,A) \qquad \text{for every } v\in W^{1,p}(\Omega;\mathbf{R}^m) \text{ and } d\in\mathbf{R}^m,$$

so that also condition (iv) is proved. ■

REMARK 4.4.4. Theorem 4.4.1 holds, with the same proof, also if instead of (4.4.2) the functional F is given by

$$F(u,A) = \begin{cases} \int_A f(x,Du)\,dx & \text{if } u\in W^{1,p}(\Omega;\mathbf{R}^m)\cap C^\infty(\Omega;\mathbf{R}^m) \\ +\infty & \text{otherwise}. \end{cases}$$

REMARK 4.4.5. Denoting by $Qf(x,z)$ the quasi-convex envelope of $f(x,z)$ with respect to z, i.e. the greatest function which is quasi-convex in z and less than or equal to $f(x,z)$, by Theorem 4.1.5 we obtain immediately

$$\int_A Qf(x,Du)\,dx \le \int_A \varphi(x,Du)\,dx \le \int_A f(x,Du)\,dx$$

for every $u\in W^{1,p}(\Omega;\mathbf{R}^m)$ and every $A\in\mathbb{A}$. Since $\varphi(x,z)$ and $Qf(x,z)$ are Carathéodory integrands, the first inequality implies (see Proposition 2.1.3)

$$Qf(x,z) \le \varphi(x,z) \qquad \text{for a.e. } x\in\Omega \text{ and every } z\in\mathbf{R}^{mn}.$$

If $f(x,z)$ is upper semicontinuous in z, the second inequality implies

$$\varphi(x,z) \le f(x,z) \qquad \text{for a.e. } x\in\Omega \text{ and every } z\in\mathbf{R}^{mn},$$

so that, since $\varphi(x,z)$ is quasi-convex in z,

$$\varphi(x,z) = Qf(x,z) \qquad \text{for a.e. } x\in\Omega \text{ and every } z\in\mathbf{R}^{mn}. \tag{4.4.14}$$

In the case $m=1$ this reduces to

$$\varphi(x,z) = f^{**}(x,z) \qquad \text{for a.e. } x\in\Omega \text{ and every } z\in\mathbf{R}^{mn},$$

where $f^{**}(x,z)$ denotes the convex envelope of $f(x,z)$ with respect to z.

But for general integrands $f(x,z)$ equality (4.4.14) does not hold, as the following example shows.

EXAMPLE 4.4.6. Let $m=1$, $n>1$, p be any number in $[1,+\infty]$, and Ω be a given bounded open subset of $\mathbf{R}^n$. Define $f:\Omega\times\mathbf{R}^n\to[0,+\infty[$ by

$$f(x,z) = \begin{cases} 0 & \text{if } z=U(x) \\ 1 & \text{otherwise} \end{cases}$$

where $U:\mathbf{R}^n\to\mathbf{R}^n$ is the function

$$U(x) = (x_n,0,\dots,0)\,.$$

Consider the functional

$$F(u,A) = \begin{cases} \displaystyle\int_A f(x,Du)\,dx & \text{if } u\in W^{1,p}(\Omega)\cap C^\infty(\Omega) \\ +\infty & \text{if } u\in W^{1,p}(\Omega)-C^\infty(\Omega) \end{cases}$$

and denote by $\Gamma F(u,A)$ its relaxation. For every $u\in C^\infty(\Omega)$ let E_u be the set

$$E_u = \{x\in\Omega : Du(x) = U(x)\};$$

since Du and U are in $C^\infty(\Omega)$, by Corollary 4.2.3 we have

$$\begin{cases} D_1D_nu = D_1U_n = 0 & \text{a.e. on } E_u \\ D_nD_1u = D_nU_1 = 1 & \text{a.e. on } E_u\,. \end{cases}$$

Therefore $D_1D_nu \neq D_nD_1u$ a.e. on E_u, which implies $\text{meas}(E_u)=0$. Then we get

$$F(u,A) = \text{meas}(A) \qquad \text{for every } A\in\mathbb{A} \text{ and } u\in C^\infty(\Omega),$$

so that

$$\Gamma F(u,A) = \text{meas}(A) \qquad \text{for every } A\in\mathbb{A} \text{ and } u\in W^{1,p}(\Omega).$$

This implies that the integrand $\varphi(x,z)$ given by Theorem 4.4.1 is $\varphi(x,z)\equiv 1$, whereas

$f^{**}(x,z) \equiv 0$.

The relaxation Theorem 4.4.1 can be also used to compute the lower semicontinuous envelope of some integral functionals of the form

$$\int_\Omega f(x,u,Du)\,dx$$

where $f(x,s,z)$ is highly discontinuous in s. Consider the following example:

$$F(u,A) = \int_A \big[f(x,Du) + \psi(u)\big]\,dx \qquad \big(u \in W^{1,p}(\Omega),\ A \in \mathbb{A}\big).$$

Here $m=1$, $p \in [1,+\infty]$, $f:\Omega \times \mathbf{R}^n \to [0,+\infty[$ is a Carathéodory integrand satisfying a growth condition of order p, and $\psi:\mathbf{R} \to \mathbf{R}$ is the function given by

$$\psi(s) = \begin{cases} 0 & \text{if s is rational} \\ 1 & \text{if s is irrational.} \end{cases}$$

THEOREM 4.4.7. *Let* $\Gamma F(u,A)$ *be the relaxed functional of* $F(u,A)$ *with respect to the sequential weak topology of* $W^{1,p}(\Omega)$ *(weak* if* $p=+\infty$*); then we have for every* $u \in W^{1,p}(\Omega)$ *and* $A \in \mathbb{A}$

$$\Gamma F(u,A) = \int_A g^{**}(x,Du)\,dx$$

where $g^{**}(x,z)$ *denotes the convex envelope (with respect to* z*) of the function*

$$g(x,z) = \begin{cases} f(x,0) & \textit{if } x \in \Omega,\ z=0 \\ 1 + f(x,z) & \textit{if } x \in \Omega,\ z \neq 0\,. \end{cases}$$

Proof. The proof consists of several steps.

Step 1. *The functional* ΓF *satisfies all conditions of the integral representation Theo-*

rem 4.3.2.

By Lemmas 4.4.2 and 4.4.3 the functional ΓF satisfies conditions (i), (ii), (iii) of Theorem 4.3.2. Since the lower semicontinuity property (v) is obvious for ΓF, it remains only to prove property (iv). For every rational number c and every $A \in \mathbb{A}$ we have

$$F(u+c,A) = F(u,A) \qquad \text{for every } u \in W^{1,p}(\Omega),$$

so that

$$\Gamma F(u+c,A) \le F(u+c,A) = F(u,A) \qquad \text{for every } u \in W^{1,p}(\Omega).$$

Hence

$$(4.4.15) \qquad \Gamma F(u+c,A) \le \Gamma F(u,A) \qquad \text{for every } u \in W^{1,p}(\Omega;\mathbf{R}^m),$$

and, by lower semicontinuity, we have that (4.4.15) holds also for every $c \in \mathbf{R}$. As in the proof of Theorem 4.4.1, the opposite inequality follows by taking $u=v+d$ and $c=-d$.

Step 2. *There exists a Carathéodory integrand* $\varphi(x,z)$, *convex in* z, *such that for every* $u \in W^{1,p}(\Omega;\mathbf{R}^m)$ *and* $A \in \mathbb{A}$

$$\Gamma F(u,A) = \int_A \varphi(x,Du)\,dx .$$

It is enough to apply Theorem 4.3.2 and Remark 4.3.3.

Step 3. *There exists* $N \in \mathbb{B}$, *with* $\text{meas}(N)=0$, *such that*

$$\varphi(x,z) \le g^{**}(x,z) \qquad \textit{for every } x \in \Omega - N \textit{ and every } z \in \mathbf{R}^n.$$

Since $\varphi(x,z)$ is convex in z, it is enough to show that there exists $N \in \mathbb{B}$, with $\text{meas}(N)=0$, such that

$$(4.4.16) \qquad \varphi(x,z) \le g(x,z) \qquad \text{for every } x \in \Omega - N \text{ and every } z \in \mathbf{R}^n.$$

Setting $u_z(x) = \langle z,x \rangle$ we have for every $z \in \mathbf{R}^n$ and $A \in \mathbb{A}$

$$\int_A \varphi(x,z)\,dx = \Gamma F(u_z,A) \le F(u_z,A) \le \int_A [1+f(x,z)]\,dx \;;$$

hence there exists $N'\in\mathbb{B}$, with $\text{meas}(N')=0$, such that

$$\varphi(x,z) \le 1+f(x,z) \qquad \text{for every } x\in\Omega-N' \text{ and every } z\in\mathbf{Q}^n.$$

Since $\varphi(x,z)$ and $f(x,z)$ are continuous in z, this implies

(4.4.17) $\quad \varphi(x,z) \le 1+f(x,z) = g(x,z) \quad$ for every $x\in\Omega-N'$ and every $z\in\mathbf{R}^n-\{0\}$.

For $z=0$ we have for every $A\in\mathbb{A}$

$$\int_A \varphi(x,0)\,dx = \Gamma F(0,A) \le F(0,A) = \int_A f(x,0)\,dx \;;$$

hence there exists $N''\in\mathbb{B}$, with $\text{meas}(N'')=0$, such that

(4.4.18) $\qquad \varphi(x,0) \le f(x,0) = g(x,0) \qquad$ for every $x\in\Omega-N''$.

By (4.4.17) and (4.4.18) we get (4.4.16) with $N=N'\cup N''$.

<u>Step 4.</u> *For every* $u\in W^{1,p}(\Omega)$ *and* $A\in\mathbb{A}$ *we have*

$$\int_A g^{**}(x,Du)\,dx \le F(u,A) \,.$$

Fix $u\in W^{1,p}(\Omega)$, $A\in\mathbb{A}$, and set

$$E = \{x\in A : u(x) \text{ is rational}\}.$$

By Proposition 4.2.2 we have $Du=0$ a.e. on E, so that

$$\int_A g^{**}(x,Du)\,dx = \int_{A-E} g^{**}(x,Du)\,dx + \int_E g^{**}(x,0)\,dx \le$$

$$\le \int_{A-E}[1+f(x,Du)]\,dx + \int_E f(x,0)\,dx = \int_{A-E}[1+f(x,Du)]\,dx + \int_E f(x,Du)\,dx =$$

$$= \int_A [f(x,Du)+\psi(u)]\,dx = F(u,A) \,.$$

<u>Step 5.</u> *For every* $u\in W^{1,p}(\Omega)$ *and* $A\in\mathbb{A}$ *we have*

$$\Gamma F(u,A) = \int_A g^{**}(x,Du)\,dx\ .$$

Fix $A \in \mathbb{A}$; by Theorem 4.1.1 the functional

$$G(u) = \int_A g^{**}(x,Du)\,dx$$

is seq. $wW^{1,p}(\Omega)$-l.s.c. (w^* if $p=+\infty$) on $W^{1,p}(\Omega)$, so that Step 4 implies

$$G(u) \le \Gamma F(u,A) \qquad \text{for every } u \in W^{1,p}(\Omega).$$

The opposite inequality follows from Step 2 and Step 3. ∎

EXAMPLE 4.4.8. Let $m=1$, $p=2$, and let

$$F(u,A) = \int_A \left[|Du|^2 + \psi(u)\right] dx$$

where $\psi(s)$ is defined as above. Then we have

$$\Gamma F(u,A) = \int_A \varphi(|Du|)\,dx$$

where $\varphi:[0,+\infty[\to[0,+\infty[$ is the function

$$\varphi(t) = \begin{cases} 2t & \text{if } 0 \le t \le 1 \\ 1 + t^2 & \text{if } t > 1\ . \end{cases}$$

4.5. Further Remarks

Many extensions of the results of this chapter have been made in the recent years, and we shall recall here some of them. As before, Ω will denote a bounded open subset of $\mathbf{R}^n$ and, as in Example 4.1.9, the $wBV(\Omega)$ convergence is defined by

$$v_h \to v \text{ in } wBV(\Omega) \iff v_h \to v \text{ in } L^1(\Omega) \text{ and } \{v_h\} \text{ is bounded in } W^{1,1}(\Omega).$$

DEFINITION 4.5.1. *Given a function* $f:\Omega\times\mathbf{R}\times\mathbf{R}^n\to[0,+\infty]$ *we say that* f *satisfies the (K)-property if for every* $\varepsilon>0$ *there exists a compact subset* K_ε *of* Ω *such that* $meas(\Omega-K_\varepsilon)<\varepsilon$ *and for every* $s\in\mathbf{R}$ *the function* $f(\cdot,s,\cdot)$ *is l.s.c. on* $K_\varepsilon\times\mathbf{R}^n$.

THEOREM 4.5.2. (See Ambrosio [11], Theorem 3.2). *Let* $f:\Omega\times\mathbf{R}\times\mathbf{R}^n\to[0,+\infty]$ *be a l.s.c. function such that*

(i) *for every* $(x,s)\in\Omega\times\mathbf{R}^n$ *the function* $f(x,s,\cdot)$ *is convex on* $\mathbf{R}^n$;

(ii) *there exists a continuous function* $z_0:\Omega\times\mathbf{R}\to\mathbf{R}^n$ *such that the function* $(x,s)\to f(x,s,z_0(x,s))$ *is continuous and finite.*

Then the functional

$$F(u) = \int_\Omega f(x,u,Du)\,dx$$

is seq. l.s.c. on $W^{1,1}(\Omega)$ *with respect to the* $wBV(\Omega)$ *convergence.*

THEOREM 4.5.3. (See Ambrosio [11], Theorem 4.14). *Let* $f:\Omega\times\mathbf{R}\times\mathbf{R}^n\to[0,+\infty]$ *be a Borel function such that*

(a) *for every* $(x,s)\in\Omega\times\mathbf{R}^n$ *the function* $f(x,s,\cdot)$ *is convex on* $\mathbf{R}^n$;

(b) *for a.e.* $x\in\Omega$ *the function* $f(x,\cdot,0)$ *is l.s.c. on* $\mathbf{R}$;

(c) *for every* $s\in\mathbf{R}$ *the function* $f(\cdot,s,\cdot)-f(\cdot,s,0)$ *is l.s.c. on* $\Omega\times\mathbf{R}^n$;

(d) *there exists a function* $\lambda:\Omega\times\mathbf{R}\to\mathbf{R}^n$ *with* $\lambda(x,s)$ *continuous in* x, *measurable in* s, *and such that*

  (i) *for every* $(x,s)\in\Omega\times\mathbf{R}$ *it is* $\lambda(x,s)\in\partial f(x,s,0)$;

(ii) *for every* $A \subset\subset \Omega$ *the function* $\sup\{|\lambda(x,\cdot)| : x \in A\}$ *is locally integrable on* $\mathbf{R}$;

(iii) *for every* $A \subset\subset \Omega$ *and every* $H \subset\subset \mathbf{R}$ *the family of functions* $\{\lambda(\cdot,s)\}_{s \in H}$ *is equicontinuous on* A.

Then the functional

$$F(u) = \int_{\Omega} f(x,u,Du)\, dx$$

is seq. l.s.c. on $W^{1,1}(\Omega)$ *with respect to the* $wBV(\Omega)$ *convergence.*

THEOREM 4.5.4. (See Ambrosio [11], Theorem 4.15). *Let* $f:\Omega \times \mathbf{R} \times \mathbf{R}^n \to [0,+\infty]$ *be a Borel function satisfying conditions* (a), (b) *of Theorem 4.5.3 and*

(c') *the function* $f(x,s,z) - f(x,s,0)$ *has the (K)-property;*

(d') *for every* $\varepsilon>0$ *there exist a compact set* $K_\varepsilon \subset \Omega$ *with* $\mathrm{meas}(\Omega - K_\varepsilon) < \varepsilon$ *and a function* $\lambda_\varepsilon : K_\varepsilon \times \mathbf{R} \to \mathbf{R}^n$ *with* $\lambda_\varepsilon(x,s)$ *continuous in* x *and measurable in* s *such that*

(i) *for every* $(x,s) \in K_\varepsilon \times \mathbf{R}$ *it is* $\lambda_\varepsilon(x,s) \in \partial f(x,s,0)$;

(ii) *the function* $\sup\{|\lambda_\varepsilon(x,\cdot)| : x \in K_\varepsilon\}$ *is locally bounded on* $\mathbf{R}$;

(iii) *for every* $H \subset\subset \mathbf{R}$ *the family of functions* $\{\lambda_\varepsilon(\cdot,s)\}_{s \in H}$ *is equicontinuous on* K_ε.

Then the functional

$$F(u) = \int_{\Omega} f(x,u,Du)\, dx$$

is seq. l.s.c. on $W^{1,1}(\Omega)$ *with respect to the* $wW^{1,1}(\Omega)$ *convergence.*

REMARK 4.5.5. Note that if $f(x,s,0)=0$ conditions (b), (d), and (d') of Theorems

4.5.3 and 4.5.4 are automatically satisfied. Moreover, by Scorza Dragoni theorem (see for instance Ekeland & Temam [148], page 235) the (K)-property is fulfilled by every normal integrand $f(x,s,z)$ and by every function of the form

$$f(x,s,z) = g(s)\, h(x,z)$$

where $g(s)$ is measurable and $h(x,z)$ is a normal integrand.

Concerning the integral representation Theorem 4.3.2, it has been generalized by Buttazzo & Dal Maso in [74] in the following way.

First of all we extend Definition 4.3.1 to functions $f:\Omega\times\mathbf{R}^m\times\mathbf{R}^{mn}\to\mathbf{R}$ by considering in (a) the inequality

$$|f(x,s,z)| \le a(x) + b(|s|^p+|z|^p) \qquad \text{for a.e. } x\in\Omega \text{ and all } (s,z)\in\mathbf{R}^m\times\mathbf{R}^{mn},$$

in (b) the inequality

$$|f(x,s,z)| \le a_r(x) \qquad \text{for a.e. } x\in\Omega \text{ and all } (s,z)\in\mathbf{R}^m\times\mathbf{R}^{mn} \text{ with } |s|,|z|\le r,$$

in (e) the inequality

$$|F(u,A)| \le \int_A \left[a(x)+b\left(|u|^p+|Du|^p\right)\right]dx \quad \text{for every } A\in\mathbb{A} \text{ and } u\in W^{1,p}(\Omega;\mathbf{R}^m),$$

and in (f) the inequality

$$|F(u,A)| \le \int_A a_r(x)\,dx \quad \text{for every } A\in\mathbb{A} \text{ and } u\in W^{1,p}(\Omega;\mathbf{R}^m) \text{ with } |u|,|Du|\le r \text{ a.e.on } A.$$

Moreover, we say that a functional $F:W^{1,p}(\Omega;\mathbf{R}^m)\times\mathbb{B}\to\mathbf{R}$ satisfies

(g) the **strong condition** (ω) if there exists a sequence of functions $\omega_k(x,r)$ integrable in x, increasing and continuous in r, and with $\omega_k(x,0)=0$, such that

$$|F(u,A) - F(v,A)| \le \int_A \omega_k(x,r)\, dx$$

whenever $k\in\mathbf{N}$, $r\ge0$, $A\in\mathbb{A}$, and $u,v\in C^1(\overline{\Omega};\mathbf{R}^m)$ with $|u|,|v|,|Du|,|Dv|\le k$ on

A and $|u-v|,|Du-Dv|\leq r$ on A.

(h) the **weak condition** (ω) if there exists a sequence of functions $\omega_k(x,r)$ integrable in x, increasing and continuous in r, and with $\omega_k(x,0)=0$, such that

$$|F(u+s,A) - F(u,A)| \leq \int_A \omega_k(x,|s|)\, dx$$

whenever $k\in\mathbf{N}$, $A\in\mathbb{A}$, $s\in\mathbf{R}^m$, and $u\in C^1(\bar{\Omega};\mathbf{R}^m)$ with $|u|,|u+s|,|Du|\leq k$ on A.

Finally, we say that a sequence $\{u_h\}$ in $W^{1,\infty}(\Omega;\mathbf{R}^m)$ is τ_∞-converging to a function $u\in W^{1,\infty}(\Omega;\mathbf{R}^m)$ if $\{u_h\}$ is bounded in $W^{1,\infty}(\Omega;\mathbf{R}^m)$, $u_h\to u$ uniformly on every compact subset of Ω, and $Du_h(x)\to Du(x)$ a.e. in Ω.

We can now state the integral representation results (see Theorems 1.7, 1.8, 1.10 of Buttazzo & Dal Maso [74]).

THEOREM 4.5.6. *Let* $p\in[1,+\infty]$; *for every functional* $F:W^{1,p}(\Omega;\mathbf{R}^m)\times\mathbb{B}\to\mathbf{R}$ *the following conditions are equivalent:*

(a) *there exists a Carathéodory integrand* $f(x,s,z)$ *satisfying a growth condition of order* p *such that*

$$F(u,B) = \int_B f(x,u,Du)\, dx \qquad \text{for every } u\in W^{1,p}(\Omega;\mathbf{R}^m) \text{ and } B\in\mathbb{B};$$

(b) F *is local on* $\mathbb{B}$, *is a measure on* $\mathbb{B}$, *satisfies a growth condition of order* p, *and the strong condition* (ω);

(c) F *is local on* $\mathbb{A}$, *is a measure on* $\mathbb{B}$, *satisfies a growth condition of order* p, *the strong condition* (ω), *and for every* $A\in\mathbb{A}$ *the function* $u\to F(u,A)$ *is strongly* $W^{1,p}(\Omega;\mathbf{R}^m)$*-l.s.c. (seq.* τ_∞*-l.s.c. in the case* $p=+\infty$*).*

THEOREM 4.5.7. *Let* $p\in[1,+\infty]$; *for every functional* $F:W^{1,p}(\Omega;\mathbf{R}^m)\times\mathbb{B}\to\mathbf{R}$ *the* *following conditions are equivalent:*

(a) *there exists a Carathéodory integrand* $f(x,s,z)$ *quasi-convex in* z *and satisfying a growth condition of order* p *such that*

$$F(u,B) = \int_B f(x,u,Du)\,dx \qquad \text{for every } u\in W^{1,p}(\Omega;\mathbf{R}^m) \text{ and } B\in\mathbb{B};$$

(b) F *is local on* $\mathbb{B}$, *is a measure on* $\mathbb{B}$, *satisfies a growth condition of order* p, *and the weak condition* (ω), *and for every* $A\in\mathbb{A}$ *the function* $u\to F(u,A)$ *is seq.* $w^*W^{1,\infty}(\Omega;\mathbf{R}^m)$*-l.s.c.;*

(c) F *is local on* $\mathbb{A}$, *is a measure on* $\mathbb{B}$, *satisfies a growth condition of order* p, *the weak condition* (ω), *and for every* $A\in\mathbb{A}$ *the function* $u\to F(u,A)$ *is seq.* $w^*W^{1,\infty}(\Omega;\mathbf{R}^m)$*-l.s.c. and strongly* $W^{1,p}(\Omega;\mathbf{R}^m)$*-l.s.c. (seq.* τ_∞*-l.s.c. in the case* $p=+\infty$*).*

The relaxation of autonomous functionals of the form

$$F(u) = \int_\Omega f(u,Du)\,dx$$

with possibly discontinuous functions $f(s,z)$, has been studied by Buttazzo & Leaci in [78] and by Ambrosio in [12], who obtained the following result.

THEOREM 4.5.8. *Let* $p\geq 1$ *and let* $f:\mathbf{R}\times\mathbf{R}^n\to\mathbf{R}$ *be a Borel function such that*

(i) *for a.e.* $s\in\mathbf{R}$ *the function* $f(s,\cdot)$ *is convex on* $\mathbf{R}^n$*;*

(ii) *for a suitable constant* $c>0$ *we have*

$$0 \leq f(s,z) \leq c(1+|s|^p+|z|^p) \qquad \text{for every } (s,z)\in\mathbf{R}\times\mathbf{R}^n.$$

Define for every $u\in W^{1,p}(\Omega)$

$$F(u) = \int_\Omega f(u,Du)\,dx$$

$$\Gamma F(u) = \Gamma\big(L^1_{loc}(\Omega)^-\big)F(u)\,.$$

Then the functional ΓF *admits on* $W^{1,p}(\Omega)$ *the integral representation*

$$\Gamma F(u) = \int_\Omega g^{**}(u,Du)\,dx$$

where $g:\mathbf{R}\times\mathbf{R}^n\to\mathbf{R}$ *is the function defined by*

$$g(s,z) = \begin{cases} f(s,z) & \text{if } z\neq 0 \\ \liminf\limits_{t\to s} f(t,0) & \text{if } z=0. \end{cases}$$

Finally, we want to recall some results about the operators $T:W^{1,p}(\Omega)\to L^q(\Omega)$ of the form

(4.5.1) $\qquad (Tu)(x) = f(x,u(x),Du(x))\,.$

If $f(x,s,z)$ is measurable in x, continuous in (s,z), and satisfies for suitable $c>0$ and $a\in L^q(\Omega)$

(4.5.2) $\qquad |f(x,s,z)| \le c\big(a(x)+|s|^{p/q}+|z|^{p/q}\big) \qquad$ for every $(x,s,z)\in\Omega\times\mathbf{R}\times\mathbf{R}^n$,

then the continuity of the operator T in (4.5.1) with respect to the strong topologies of $W^{1,p}(\Omega)$ and $L^q(\Omega)$ follows from the Lebesgue dominated convergence theorem (see for instance Vainberg [282]). If $f(x,s,z)$ is only a Borel function which satisfies (4.5.2), then the continuity of T (with respect to the strong topologies of $W^{1,p}(\Omega)$ and $L^q(\Omega)$) still holds, provided the following weaker property is satisfied (see Buttazzo & Leaci [77] and Ambrosio & Buttazzo & Leaci [15]):

(4.5.3) (i) *for every* $x\in\Omega$ *and* $s\in\mathbf{R}$ *the function* $f(x,\cdot,\cdot)$ *is continuous at* $(s,0)$;

(ii) *for every* $\varepsilon>0$ *there exist compact sets* $K_\varepsilon\subset\Omega$ *and* $H_\varepsilon\subset[-1/\varepsilon,1/\varepsilon]$ *such that* $meas(\Omega-K_\varepsilon)<\varepsilon$, $meas([-1/\varepsilon,1/\varepsilon]-H_\varepsilon)<\varepsilon$, *and the function* f

restricted to $H_\varepsilon \times K_\varepsilon \times \mathbf{R}^n$ *is continuous.*

For instance, by using the Scorza-Dragoni theorem, property (4.5.3) is fulfilled in the following situations:

-) $f(x,s,z)$ measurable in x and continuous in (s,z);

-) $f(x,s,z)$ measurable in s and continuous in (x,z);

-) $f(x,s,z)=g(x)h(s,z)$ with $g(x)$ measurable and $h(s,z)$ measurable in s and continuous in z;

-) $f(x,s,z)=g(s)h(x,z)$ with $g(s)$ measurable and $h(x,z)$ measurable in x and continuous in z.

When $p=q=2$ and $f(x,s,z)$ is linear in z, that is

$$f(x,s,z) = \sum_{i=1}^{n} a_i(x,s)\, z_i \ ,$$

then the (sequential) continuity of T has been proved also if $W^{1,2}(\Omega)$ and $L^2(\Omega)$ are endowed with their weak topologies (see Boccardo & Buttazzo [51]), provided a property ananogous to the one of Definition 4.5.1 is satisfied, that is

(4.5.4) *for all* $\varepsilon>0$ *there exists a compact set* $K_\varepsilon \subset \Omega$ *such that* $\text{meas}(\Omega - K_\varepsilon)<\varepsilon$, *and for all* $R>0$ *the family of functions* $\{a(\cdot,s)\}_{|s|\le R}$ *is equicontinuous on* K_ε.

This allows us to prove existence results for quasilinear elliptic equations of the form

$$\begin{cases} -D_i\big(a_{ij}(x,u)D_ju\big) = g & \text{on } \Omega \\ u \in H_0^1(\Omega) \end{cases}$$

where $g \in H^{-1}(\Omega)$ and $\{a_{ij}(x,s)\}$ is a $n\times n$ matrix satisfying (4.5.4) and the usual ellipticity and boundedness conditions

$$\begin{cases} \lambda |z|^2 \le a_{ij}(x,s) z_i z_j \\ |a_{ij}(x,s)| \le \Lambda \, . \end{cases} \qquad (0<\lambda\le\Lambda)$$

CHAPTER 5

Relaxation Problems in Optimal Control Theory

This chapter is devoted to the study of relaxation in the framework of optimal control problems. In this kind of problems we have two variables which play an important role: the first one (called state variable) describes the state of the physical system, and the second (the control variable) describes how an exterior operator may act on the system. The two variables are related by the physical laws which govern the system (often differential equations or inequalities), and the control variable has to be choosen in order to minimize a given functional (called cost functional).

We may refer the reader interested in an exhaustive presentation of optimal control theory to the books of Berkovitz [46], Lee & Marcus [191], Cesari [92], Clarke [100], where a lot of examples from biology, chemistry, engineering are presented. In many cases of systems governed by ordinary differential equations, one has to deal with optimal control problems of the form

$$\min \left\{ \int_0^T f(t,y,u)\, dt \ : \ y'=g(t,y,u)\ ,\ y(0)=y_0 \right\}$$

where y is the state variable (belonging to some suitable Sobolev spaces), u is the control variable (belonging to some L^p spaces), and f, g are two given functions.

In order to develop the relaxation theory for such kind of problems with two variables, we shall consider multiple Γ-operators in an abstract framework (Sections 5.1 and 5.2), and we shall apply the results obtained to control problems governed by ordinary and partial differential equations (Sections 5.3 and 5.4).

5.1. Multiple Γ-Operators

In this section we shall generalize the construction of the relaxed functional $\Gamma(X^-)F$ given in Section 1.3 to the case of functions F defined on the product $X_1 \times X_2$ of two topological spaces. For our purposes it is enough to consider only the sequential definition of multiple Γ-operators, and for simplicity we shall restrict ourselves to them, referring the interested reader to Attouch [21], Buttazzo [65], De Giorgi [128] for a more systematic presentation.

Let X_1, X_2 be topological spaces, let $F: X_1 \times X_2 \to \overline{\mathbf{R}}$ be a function, and let $(x_1,x_2) \in X_1 \times X_2$. We indicate by $Z(+)$ the "sup" operator and by $Z(-)$ the "inf" operator. For $i=1,2$ let S_i be the set of all sequences in X_i converging to x_i, and let α_i be one of the signs + or −. We define

$$\Gamma_{seq}(X_1^{\alpha_1}, X_2^{\alpha_2})\, F(x_1,x_2) \;=\; \underset{(x_1^h)\in S_1}{Z(\alpha_1)}\; \underset{(x_2^h)\in S_2}{Z(\alpha_2)}\; \underset{k\in \mathbf{N}}{Z(-\alpha_1)}\; \underset{h\geq k}{Z(\alpha_1)}\; F(x_1^h, x_2^h)\,.$$

For example we have

$$\Gamma_{seq}(X_1^-, X_2^-)\, F(x_1,x_2) \;=\; \inf_{x_1^h \to x_1}\; \inf_{x_2^h \to x_2}\; \liminf_{h\to+\infty} F(x_1^h, x_2^h)\;,$$

$$\Gamma_{seq}(X_1^-, X_2^+)\, F(x_1,x_2) \;=\; \inf_{x_1^h \to x_1}\; \sup_{x_2^h \to x_2}\; \liminf_{h\to+\infty} F(x_1^h, x_2^h)\;.$$

When a Γ_{seq}-limit is independent of the sign + or − associated to one of the spaces, then this sign is omitted. For instance, if

$$\Gamma_{seq}(X_1^-, X_2^+)\, F(x_1,x_2) \;=\; \Gamma_{seq}(X_1^+, X_2^+)\, F(x_1,x_2)\;,$$

then their common value will indicated by

$$\Gamma_{seq}(X_1, X_2^+)\, F(x_1,x_2)\,.$$

REMARK 5.1.1. It is immediate to see that

$$\Gamma_{seq}(X_1^{\alpha_1},X_2^{\alpha_2})\,F(x_1,x_2) \;=\; -\Gamma_{seq}(X_1^{-\alpha_1},X_2^{-\alpha_2})\,(-F)(x_1,x_2)\;;$$

moreover, if $X_1 \times X_2$ is metrizable, or if conditions of Proposition 1.3.5 are satisfied, we have

$$\Gamma_{seq}(X_1^-,X_2^-)\,F(x_1,x_2) \;=\; \Gamma\big((X_1\times X_2)^-\big)\,F(x_1,x_2)\;.$$

PROPOSITION 5.1.2. *The following properties hold:*

(i) $\inf_{X_1\times X_2} F \;=\; \inf_{X_1\times X_2} \Gamma_{seq}(X_1^-,X_2^-)\,F\;;$

(ii) *for every sequentially continuous function* $G:X_1\times X_2\to\mathbf{R}$ *we have*

$$\Gamma_{seq}(X_1^-,X_2^-)[G+F] \;=\; G+\Gamma_{seq}(X_1^-,X_2^-)\,F\;.$$

(iii) *if* (x^h,x^h) *is a minimizing sequence for* F *which converges to some point* $(x_1,x_2)\in X_1\times X_2$, *then* (x_1,x_2) *is a minimum point for* $\Gamma_{seq}(X_1^-,X_2^-)F$ *on* $X_1\times X_2$.

Proof. It is similar to the one of Proposition 1.3.1. ∎

In general Γ-operators don't commute with the sum; nevertheless, some inequalities are guaranteed, as the following proposition shows.

PROPOSITION 5.1.3. *Let* $F,G:X_1\times X_2\to\overline{\mathbf{R}}$; *then, if all the sums are well-defined,*

$$\Gamma_{seq}(X_1^-,X_2^-)F+\Gamma_{seq}(X_1^-,X_2^-)G \;\le\; \Gamma_{seq}(X_1^-,X_2^-)[F+G] \;\le$$

$$\le\; \Gamma_{seq}(X_1^-,X_2^+)F+\Gamma_{seq}(X_1^+,X_2^-)G\;.$$

Proof. It is enough to apply repeatedly the inequalities

$$\inf_{i\in I} a_i + \inf_{i\in I} b_i \leq \inf_{i\in I} (a_i+b_i) \leq \inf_{i\in I} a_i + \sup_{i\in I} b_i$$

valid for every $\{a_i\}_{i\in I}$ and $\{b_i\}_{i\in I}$ families in $\overline{\mathbf{R}}$. ■

REMARK 5.1.4. Other inequalities can be obtained in a similar way; for instance

$$\Gamma_{seq}(X_1^-,X_2^-)F + \Gamma_{seq}(X_1^-,X_2^+)G \leq \Gamma_{seq}(X_1^-,X_2^+)[F+G] \leq$$

$$\leq \Gamma_{seq}(X_1^-,X_2^+)F + \Gamma_{seq}(X_1^+,X_2^+)G ;$$

$$\Gamma_{seq}(X_1^-,X_2^-)[F+G] \leq \Gamma_{seq}(X_1^-,X_2^-)F + \Gamma_{seq}(X_1^+,X_2^+)G .$$

For our purposes, the following consequence of Proposition 5.1.3 will be useful.

COROLLARY 5.1.5. *Assume that for a given point* $(x_1,x_2)\in X_1\times X_2$ *there exist*

$$\Gamma_{seq}(X_1^-,X_2) F(x_1,x_2) \qquad \textit{and} \qquad \Gamma_{seq}(X_1,X_2^-) G(x_1,x_2) .$$

Then we have

$$\Gamma_{seq}(X_1^-,X_2^-) [F+G](x_1,x_2) = \Gamma_{seq}(X_1^-,X_2) F(x_1,x_2) + \Gamma_{seq}(X_1,X_2^-) G(x_1,x_2) .$$

5.2. The Abstract Framework

An optimal control problem can be formulated in the following way. Let Y (the space of state variables) and U (the space of controls) be two topological spaces, let

$J:U\times Y\to[0,+\infty]$ be a function (the cost function), and let Λ be a subset of $U\times Y$ (the set of admissible pairs, often determined by differential equations or differential inclusions, constraints, etc...). The problem of optimal control is then

$$\min\{J(u,y)\ :\ (u,y)\in\Lambda\},$$

or equivalently, setting

$$\chi_\Lambda(u,y) = \begin{cases} 0 & \text{if } (u,y)\in\Lambda \\ +\infty & \text{otherwise}, \end{cases}$$

the problem becomes

$$\min\{[J+\chi_\Lambda](u,y)\ :\ (u,y)\in U\times Y\}.$$

Denote for simplicity by F the functional $J+\chi_\Lambda$ and by ΓF its relaxation

$$\Gamma F(u,y) = \Gamma_{seq}(U^-,Y^-)\,F(u,y)\,.$$

We shall be concerned with the problem of characterizing explicitely the functional ΓF in terms of the cost function J and the admissible set Λ. To do that it may be useful to introduce a new "auxiliary variable" whose effect on the relaxation is given by the following proposition.

PROPOSITION 5.2.1. *Let* V *be another topological space, and let* $\vartheta:U\times Y\to V$ *be a mapping such that the following compactness condition holds:*

(5.2.1) *for every sequence* (u_h,y_h) *converging in* $U\times Y$, *with* $F(u_h,y_h)$ *bounded, the sequence* $\vartheta(u_h y_h)$ *is compact in* V *(in the sense that it admits a subsequence converging in* V*).*

Then, setting

$$G(u,v,y) = \begin{cases} F(u,y) & \textit{if } v=\vartheta(u,y) \\ +\infty & \textit{otherwise}, \end{cases}$$

we have for every $(u,y)\in U\times Y$

$$\Gamma F(u,y) = \inf\{\Gamma_{seq}((U\times V)^-,Y^-)\,G(u,v,y) \;:\; v\in V\}.$$

Proof. Let $(u,v,y)\in U\times V\times Y$ and let $u_h\to u$, $v_h\to v$, $y_h\to y$ in U, V, Y respectively. Then we have

$$F(u_h,y_h) \le G(u_h,v_h,y_h) \qquad \text{for every } h\in\mathbf{N},$$

so that

$$\Gamma F(u,y) \le \Gamma_{seq}((U\times V)^-,Y^-)\,G(u,v,y)\,,$$

and, since $v\in V$ is arbitrary,

$$\Gamma F(u,y) \le \inf\{\Gamma_{seq}((U\times V)^-,Y^-)\,G(u,v,y) \;:\; v\in V\}.$$

In order to prove the opposite inequality, fix $(u,v)\in U\times Y$ and $u_h\to u$ in U, $y_h\to y$ in Y; we have to show that

$$\text{(5.2.2)} \qquad \liminf_{h\to+\infty} F(u_h,y_h) \ge \inf\left\{\Gamma_{seq}\left((U\times V)^-,Y^-\right) G(u,v,y) \;:\; v\in V\right\}.$$

It is not restrictive to assume that the liminf at the left-hand side of (5.2.2) is a limit and that $F(u_h,y_h)$ is bounded; therefore by property (5.2.1), possibly passing to a subsequence, we have that $v_h=\vartheta(u_h,y_h)$ converges in V to some $v\in V$. Then

$$\Gamma_{seq}\left((U\times V)^-,Y^-\right) G(u,v,y) \le \liminf_{h\to+\infty} G(u_h,v_h,y_h) = \liminf_{h\to+\infty} F(u_h,y_h)\,,$$

and so (5.2.2) is proved. ∎

PROPOSITION 5.2.2. *Assume the cost functional* J *satisfies the following condition:*

$$\text{(5.2.3)} \qquad J(u,y) \le J(u,z) + \omega(y,z)\,H(u,z) \qquad \forall u\in U,\ \forall y,z\in Y$$

for suitable functions $\omega:Y\times Y\to[0,+\infty]$ *and* $H:U\times Y\to[0,+\infty]$ *such that*

(5.2.4) $$y_h \to y \text{ in } Y \ \Rightarrow\ \lim_{h\to+\infty} \omega(y,y_h) = \lim_{h\to+\infty} \omega(y_h,y) = 0 \ ;$$

(5.2.5) (u_h,y_h) *convergent in* $U\times Y$ *and* $J(u_h,y_h)$ *bounded* $\Rightarrow$ $H(u_h,y_h)$ *bounded.*

Then, for every $(u,y)\in U\times Y$ *there exists* $\Gamma_{seq}(U^-,Y)J(u,y)$ *and we have*

$$\Gamma_{seq}(U^-,Y)J(u,y) = \Gamma_{seq}(U^-,\delta_Y)J(u,y)$$

where δ_Y *denotes the discrete topology on* Y.

Proof. Let $(u,y)\in U\times Y$; to prove the inequality

$$\Gamma_{seq}(U^-,Y^-)J(u,y) \geq \Gamma_{seq}(U^-,\delta_Y)J(u,y)$$

it is enough to show that

(5.2.6) $$\liminf_{h\to+\infty} J(u_h,y_h) \geq \liminf_{h\to+\infty} J(u_h,y)$$

whenever $(u_h,y_h)\to(u,y)$ in $U\times Y$. Possibly passing to subsequences, in the right-hand side of (5.2.6) we may assume that the liminf is a limit, and that $J(u_h,y_h)$ is bounded. Then, by (5.2.3) we get

$$J(u_h,y) \leq J(u_h,y_h) + \omega(y,y_h)\, H(u_h,y_h) \qquad \forall h\in \mathbf{N},$$

and by (5.2.4) and (5.2.5) inequality (5.2.6) follows.

In order to prove the inequality

$$\Gamma_{seq}(U^-,Y^+)J(u,y) \leq \Gamma_{seq}(U^-,\delta_Y)J(u,y)$$

it is enough to show that

(5.2.7) $$\liminf_{h\to+\infty} J(u_h,y_h) \leq \liminf_{h\to+\infty} J(u_h,y)$$

whenever $(u_h,y_h)\to(u,y)$ in $U\times Y$. Again, possibly passing to subsequences, in the right-hand side of (5.2.7) we may assume that the liminf is a limit, and that $J(u_h,y)$ is bounded. Then, by (5.2.3)

$$J(u_h,y_h) \leq J(u_h,y) + \omega(y_h,y)\, H(u_h,y) \qquad \forall h\in \mathbf{N},$$

and by (5.2.4) and (5.2.5), this implies (5.2.7). ■

5.3. Problems with Ordinary State Equations

In this section we apply the abstract results of previous sections to characterize the relaxed functional related to some optimal control problems governed by ordinary differential state equations.

Let k, m, n be positive integers and let $T>0$ be fixed; denote by Y the space $W^{1,1}(0,T;\mathbf{R}^n)$ endowed with the $L^\infty(0,T;\mathbf{R}^n)$ topology, and by U the space $L^1(0,T;\mathbf{R}^m)$ endowed with its weak topology: the spaces Y and U will be the space of state variables and the space of controls respectively. We consider a cost functional of the form

$$J(u,y) = \int_0^T f(t,y,u)\,dt \qquad (u\in U,\ y\in Y) \tag{5.3.1}$$

where $f:[0,T]\times\mathbf{R}^n\times\mathbf{R}^m\to[0,+\infty]$ is a Borel function, and a state equation of the form

$$\begin{cases} y' = A(t,y) + B(t,y)\,b(t,u) & \text{a.e. in } [0,T] \\ y(0)\in K \end{cases} \tag{5.3.2}$$

where $A:[0,T]\times\mathbf{R}^n\to\mathbf{R}^n$, $B:[0,T]\times\mathbf{R}^n\to\mathbf{R}^{kn}$, $b:[0,T]\times\mathbf{R}^m\to\mathbf{R}^k$ are Borel functions, and K is a closed subset of $\mathbf{R}^n$. The admissible set Λ is then defined by

$$\Lambda = \{(u,y)\in U\times Y : (u,y) \text{ satisfies } (5.3.2)\}.$$

Note that all possible constraints on the controls, like $u(t)\in\Xi(t)$, are incorporated in the cost functional, which may assume the value $+\infty$.

About the function f in (5.3.1) we assume that:

(5.3.3) for every $r \geq 0$, $t \in [0,T]$, $u \in \mathbf{R}^m$, $y,z \in \mathbf{R}^n$ with $|y|,|z| \leq r$ we have

$$f(t,y,u) \leq f(t,z,u) + \rho_r(t,|y-z|) + \sigma_r(|y-z|)\, f(t,z,u)$$

where $\rho_r:[0,T]\times[0,+\infty[\to[0,+\infty[$ and $\sigma_r:[0,+\infty[\to[0,+\infty[$ are such that $\rho_r(t,s)$ is integrable in t, $\rho_r(t,s)$ ans $\sigma_r(s)$ are increasing and continuous in s, and $\rho_r(t,0)=\sigma_r(0)=0$;

(5.3.4) there exist $a \in L^1(0,T)$ and $\psi:[0,+\infty[\to[0,+\infty[$ (which can be taken increasing and convex) such that

$$\lim_{s\to+\infty} \frac{\psi(s)}{s} = +\infty ,$$

$$\psi(|b(t,u)| + |u|) - a(t) \leq f(t,0,u) \qquad \forall t \in [0,T],\ \forall u \in \mathbf{R}^m;$$

(5.3.5) there exists $u_0 \in L^1(0,T;\mathbf{R}^m)$ such that the function $f(t,0,u_0(t))$ is integrable on $[0,T]$.

Denote by F the functional $F=J+\chi_\Lambda$; the following lemma shows that the method of the "auxiliary variable" can be applied to F.

LEMMA 5.3.1. *Let* V *be the space* $L^1(0,T;\mathbf{R}^k)$ *endowed with its weak topology, and let* $\vartheta:U\times Y\to V$ *be given by*

$$[\vartheta(u,y)](t) = \begin{cases} b(t,u(t)) & \textit{if } b(t,u(t)) \textit{ is integrable on } [0,T] \\ 0 & \textit{otherwise.} \end{cases}$$

Then, the compactness property (5.2.1) holds for the functional F.

Proof. Let (u_h,y_h) be a sequence in $U\times Y$ converging to (u,y) such that

(5.3.6) $$F(u_h,y_h) \leq c \qquad \forall h \in \mathbf{N},$$

and let $v_h(t)=b(t,u_h(t))$. Set for simplicity

$$\varepsilon_h = \|y_h - y\|_{L^\infty(0,T;\mathbf{R}^n)} \qquad \text{and} \qquad r = \sup_{h\in\mathbf{N}} \|y_h\|_{L^\infty(0,T;\mathbf{R}^n)} .$$

Then, by (5.3.3), (5.3.4) and (5.3.6) we get

$$\int_0^T \left[\psi(|v_h|) - a(t)\right] dt \le \int_0^T f(t,0,u_h)\, dt \le$$

$$\le \int_0^T \left[f(t,y_h,u_h) + \rho_r(t,\varepsilon_h) + \sigma_h(\varepsilon_h)\, f(t,y_h,u_h)\right] dt \le$$

$$\le c\big(1 + \sigma_r(\varepsilon_h)\big) + \int_0^T \rho_r(t,\varepsilon_h)\, dt .$$

Hence the sequence $\psi(|v_h|)$ is bounded in $L^1(0,T)$ and, since ψ has a superlinear growth, by the Dunford-Pettis Theorem 1.2.8 this implies that $\{v_h\}$ is compact in V. ■

By Proposition 5.2.1 the study of relaxation of the functional F is reduced to the relaxation in $(U\times V)\times Y$ of the functional

$$G(u,v,y) = \Phi(u,v,y) + \chi_\Theta(u,v,y)$$

where

$$\Phi(u,v,y) = \int_0^T \varphi(t,y,u,v)\, dt$$

$$\varphi(t,y,u,v) = f(t,y,u) + \chi_{\{v = b(t,u)\}}$$

$$\Theta = \{(u,v,y)\in U\times V\times Y \ : \ y' = A(t,y) + B(t,y)\, v,\ \ y(0)\in K\}.$$

Now, in order to apply Corollary 5.1.5 we have to compute separately

$$\Gamma_{seq}((U\times V)^-,Y)\, \Phi(u,v,y) \qquad \text{and} \qquad \Gamma_{seq}(U\times V,Y^-)\, \chi_\Theta(u,v,y).$$

LEMMA 5.3.2. *For every* $(u,v,y) \in U \times V \times Y$ *we have*

$$\Gamma_{seq}((U\times V)^-,Y)\,\Phi(u,v,y) = \int_0^T \varphi^{**}(t,y,u,v)\,dt$$

where $\varphi^{**}(t,y,u,v)$ *denotes the convex l.s.c. envelope of* $\varphi(t,y,u,v)$ *with respect to* (u,v).

Proof. By property (5.3.3) we obtain

$$\Phi(u,v,y) \le \Phi(u,v,z) + \int_0^T \rho_r(t,\|y-z\|_{L^\infty})\,dt + \sigma_r(\|y-z\|_{L^\infty})\,\Phi(u,v,z)$$

whenever $r\ge 0$, $u\in U$, $v\in V$, $y,z\in Y$ with $\|y\|_{L^\infty},\|z\|_{L^\infty}\le r$. Hence, conditions (5.2.3), (5.2.4), (5.2.5) of Proposition 5.2.2 are satisfied, with

$$\omega(y,z) = \int_0^T \rho_r(t,\|y-z\|_{L^\infty})\,dt + \sigma_r(\|y-z\|_{L^\infty}) \qquad \text{(with } \|y\|_{L^\infty},\|z\|_{L^\infty}\le r\text{)}$$

$$H(u,v,z) = 1 + \Phi(u,v,z),$$

and so for every $(u,v,y)\in U\times V\times Y$ we have

$$\Gamma_{seq}((U\times V)^-,Y)\,\Phi(u,v,y) = \Gamma_{seq}((U\times V)^-,\delta_Y)\,\Phi(u,v,y).$$

Moreover, by Proposition 1.3.5 and Remark 1.3.6, thanks to Proposition 1.2.9 and condition (5.3.4), the quantity $\Gamma_{seq}((U\times V)^-,\delta_Y)\Phi(u,v,y)$ coincides with the relaxation in (u,v) of the functional

$$\int_0^T \varphi(t,y(t),u,v)\,dt$$

with respect to the (sequential) weak $L^1(0,T;\mathbf{R}^{m+k})$ topology. By Theorem 2.6.4 and Remark 2.6.5 this relaxed functional can be computed, and finally we get

$$\Gamma_{seq}((U\times V)^-,Y)\,\Phi(u,v,y) \;=\; \int_0^T \varphi^{**}(t,y,u,v)\,dt\;. \blacksquare$$

In order to compute the functional $\Gamma_{seq}(U\times V,Y^-)\chi_\Theta(u,v,y)$ we prove now a result concerning the continuous dependence of the solutions of an ordinary differential equation of the form

$$\begin{cases} y' = g(t,y) \\ y(0) = \eta \end{cases}$$

on the initial datum η and on the right-hand side g. For every $h\in\mathbf{N}\cup\{\infty\}$ let $g_h:[0,T]\times\mathbf{R}^n\to\mathbf{R}^n$ be Carathéodory functions satisfying the following properties:

(5.3.7) for every $h\in\mathbf{N}\cup\{\infty\}$, $r\geq 0$, $t\in[0,T]$, $y,z\in\mathbf{R}^n$ with $|y|,|z|\leq r$

$$|g_h(t,y)-g_h(t,z)| \;\leq\; L_h(t,r)\,|y-z|$$

$$\text{with}\qquad \sup_{h\in\mathbf{N}\cup\{\infty\}} \int_0^T L_h(t,r)\,dt \;=\; L(r) \;<\; +\infty\,;$$

(5.3.8) for every $h\in\mathbf{N}\cup\{\infty\}$, $r\geq 0$, $t\in[0,T]$, $y\in\mathbf{R}^n$ with $|y|\leq r$

$$|g_h(t,y)| \;\leq\; M_h(t,r)$$

$$\text{with}\qquad \sup_{h\in\mathbf{N}\cup\{\infty\}} \int_0^T M_h(t,r)\,dt \;=\; M(r) \;<\; +\infty\,;$$

(5.3.9) for every $y\in\mathbf{R}^n$

$$\lim_{h\to\infty} \Big\{ \sup_{0\leq t\leq T} \Big| \int_0^t [g_h(s,y)-g_\infty(s,y)]\,ds \Big| \Big\} \;=\; 0\,.$$

PROPOSITION 5.3.3. *Let* $g_h(t,y)$ *satisfy (5.3.7), (5.3.8), (5.3.9), and let* $\eta_h\to\eta_\infty$

in $\mathbf{R}^n$. *Then, the following statements (i) and (ii) hold.*

(i) *Let* $y_h \in W^{1,1}(0,T;\mathbf{R}^n)$ *be the solutions of*

$$\begin{cases} y' = g_h(t,y) & \textit{on } [0,T] \\ y(0) = \eta_h\,; \end{cases}$$

if $y_h \to y_\infty$ *in* $L^\infty(0,T;\mathbf{R}^n)$, *then* y_∞ *is in* $W^{1,1}(0,T;\mathbf{R}^n)$ *and solves the problem*

$$\begin{cases} y' = g_\infty(t,y) & \textit{on } [0,T] \\ y(0) = \eta_\infty\,. \end{cases}$$

(ii) *Let* $y_\infty \in W^{1,1}(0,T;\mathbf{R}^n)$ *be the solution of*

$$\begin{cases} y' = g_\infty(t,y) & \textit{on } [0,T] \\ y(0) = \eta_\infty\,; \end{cases}$$

then, for every h *large enough, there exists* $y_h \in W^{1,1}(0,T;\mathbf{R}^n)$ *solution of*

$$\begin{cases} y' = g_h(t,y) & \textit{on } [0,T] \\ y(0) = \eta_h \end{cases}$$

and $y_h \to y_\infty$ *in* $L^\infty(0,T;\mathbf{R}^n)$.

Proof. We first prove that for every $y \in C([0,T];\mathbf{R}^n)$

$$\text{(5.3.10)} \qquad \lim_{h\to\infty} \Big\{ \sup_{0\le t\le T} \Big| \int_0^t [g_h(s,y(s)) - g_\infty(s,y(s))]\, ds \Big| \Big\} = 0\,.$$

If $y(t)$ is a piecewise constant function, then (5.3.10) follows from hypothesis (5.3.9). If $y \in C([0,T];\mathbf{R}^n)$, let $r > \|y\|_{L^\infty}$ and, for every $\varepsilon > 0$, let y_ε be a piecewise constant function such that

$$\text{(5.3.11)} \qquad \|y_\varepsilon\|_{L^\infty(0,T;\mathbf{R}^n)} < r \qquad \text{and} \qquad \|y_\varepsilon - y\|_{L^\infty(0,T;\mathbf{R}^n)} < \varepsilon\,.$$

Then, by using (5.3.7) and (5.3.11) we have

$$\left|\int_0^t [g_h(s,y(s)) - g_\infty(s,y(s))]\, ds\right| \le$$

$$\le \left|\int_0^t [g_h(s,y_\varepsilon(s)) - g_\infty(s,y_\varepsilon(s))]\, ds\right| + 2\varepsilon L(r) ,$$

which implies (5.3.10).

We now prove (i). Let $y_h \in W^{1,1}(0,T;\mathbf{R}^n)$ be solutions of

$$\begin{cases} y' = g_h(t,y) & \text{on } [0,T] \\ y(0) = \eta_h , \end{cases}$$

assume $y_h \to y_\infty$ in $L^\infty(0,T;\mathbf{R}^n)$, and let $r>0$ be such that

$$\|y_\infty\|_{L^\infty(0,T;\mathbf{R}^n)} < r .$$

Then, for h large enough, we have

$$\left|\int_0^t [g_h(s,y_h(s)) - g_\infty(s,y_\infty(s))]\, ds\right| \le$$

$$\le \int_0^t |g_h(s,y_h(s)) - g_h(s,y_\infty(s))|\, ds + \left|\int_0^t [g_h(s,y_\infty(s)) - g_\infty(s,y_\infty(s))]\, ds\right| \le$$

$$\le L(r)\, \|y_h - y_\infty\|_{L^\infty(0,T;\mathbf{R}^n)} + \left|\int_0^t [g_h(s,y_\infty(s)) - g_\infty(s,y_\infty(s))]\, ds\right| ,$$

and so, by (5.3.10), for every $t \in [0,T]$

(5.3.12) $$\lim_{h\to\infty} \int_0^t g_h(s,y_h(s))\, ds = \int_0^t g_\infty(s,y_\infty(s))\, ds .$$

For $y_h(t)$ the formula

$$y_h(t) = \eta_h + \int_0^t g_h(s,y_h(s))\, ds$$

holds, so that, passing to the limit as $h \to \infty$, and using (5.3.12) we obtain

$$y_\infty(t) = \eta_\infty + \int_0^t g_\infty(s,y_\infty(s))\, ds \ ,$$

that is y_∞ is a solution of

$$\begin{cases} y' = g_\infty(t,y) & \text{on } [0,T] \\ y(0) = \eta_\infty . \end{cases}$$

We now prove (ii). Let y_∞ be a solution of

$$\begin{cases} y' = g_\infty(t,y) & \text{on } [0,T] \\ y(0) = \eta_\infty \end{cases}$$

and let $y_h:[0,\tau_h[\to \mathbf{R}^n$ be the maximal solution of

$$\begin{cases} y' = g_h(t,y) \\ y(0) = \eta_h . \end{cases}$$

If we set

$$r_0 = \max_{0 \le t \le T} |y_\infty(t)|$$

$$r = r_0 + 1$$

$$t_h = \sup \left\{ t \in [0,\tau_h[\cap [0,T] \ : \ |y_h(s)| \le r \ \ \forall s \in [0,T] \right\} ,$$

we have immediately

(5.3.13) $\qquad t_h < T \ \Rightarrow \ |y_h(t_h)| = r .$

We show that for h large enough y_h is a global solution on $[0,T]$, that is $t_h = T$.

Setting

$$\varepsilon_h = \sup_{0\le t\le T} \left|\int_0^t \left[g_h(s,y_\infty(s)) - g_\infty(s,y_\infty(s))\right] ds\right| ,$$

by (5.3.10) we have $\varepsilon_h \to 0$; moreover, for every $t \le t_h$

$$|y_h(t) - y_\infty(t)| \le |\eta_h - \eta_\infty| + \left|\int_0^t \left[g_h(s,y_h(s)) - g_\infty(s,y_\infty(s))\right] ds\right| \le$$

$$\le |\eta_h - \eta_\infty| + \int_0^t |g_h(s,y_h(s)) - g_h(s,y_\infty(s))|\, ds + \left|\int_0^t \left[g_h(s,y_\infty(s)) - g_\infty(s,y_\infty(s))\right] ds\right| \le$$

$$\le |\eta_h - \eta_\infty| + \int_0^t L_h(s,r)\, |y_h(s) - y_\infty(s)|\, ds + \varepsilon_h .$$

By Gronwall's lemma, for every $t \le t_h$ we get

$$|y_h(t) - y_\infty(t)| \le \left(\varepsilon_h + |\eta_h - \eta_\infty|\right) e^{L(r)} , \tag{5.3.14}$$

and so, if h is large enough to have

$$\left(\varepsilon_h + |\eta_h - \eta_\infty|\right) e^{L(r)} \le \frac{1}{2} ,$$

we obtain

$$|y_h(t_h)| \le r_0 + \frac{1}{2} .$$

By (5.3.13) this implies $t_h = T$ and, by (5.3.14), $y_h \to y_\infty$ in $L^\infty(0,T;\mathbf{R}^n)$. ∎

REMARK 5.3.4. The convergence condition (5.3.9) is implied by the following one:

$$g_h(\cdot,y) \to g_\infty(\cdot,y) \text{ weakly in } L^1(0,T;\mathbf{R}^n) \text{ for every } y \in \mathbf{R}^n.$$

In fact, it is well-known that if a sequence w_h converges to zero weakly in the space $L^1(0,T;\mathbf{R}^n)$, their primitives W_h (with $W_h(0)=0$) converge to zero uniformly on $[0,T]$.

We are now in a position to compute the functional $\Gamma_{seq}(U\times V, Y^-)\chi_\Theta$. On the functions $A(t,y)$ and $B(t,y)$ we assume that:

(5.3.15) for every $r\ge 0$, $t\in[0,T]$, $y,z\in \mathbf{R}^n$ with $|y|,|z|\le r$

$$|A(t,y) - A(t,z)| \le M(t,r)\,|y - z|$$

$$|B(t,y) - B(t,z)| \le N(r)\,|y - z|$$

with $M(\cdot,r)\in L^1(0,T)$ and $N(r)<+\infty$;

(5.3.16) for every $r\ge 0$, $t\in[0,T]$, $y\in \mathbf{R}^n$ with $|y|\le r$

$$|A(t,y)| \le P(t,r)$$

$$|B(t,y)| \le Q(r)$$

with $P(\cdot,r)\in L^1(0,T)$ and $Q(r)<+\infty$.

Then, the following result holds.

PROPOSITION 5.3.5. *For every* $(u,v,y)\in U\times V\times Y$ *we have*

$$\Gamma_{seq}(U\times V, Y^-)\chi_\Theta(u,v,y) = \chi_\Theta(u,v,y).$$

Proof. Fix $(u,v,y)\in U\times V\times Y$; we have to prove that

(5.3.17) $\quad (u_h,v_h,y_h)\to(u,v,y)$ with $(u_h,v_h,y_h)\in\Theta$ for infinitely many $h\in\mathbf{N}$

$\Rightarrow \ (u,v,y)\in\Theta$;

(5.3.18) $\quad (u,v,y)\in\Theta \ \Rightarrow\ \forall(u_h,v_h)\to(u,v)\ \exists y_h\to y$ with $(u_h,v_h,y_h)\in\Theta$ for every $h\in\mathbf{N}$ large enough.

Let us prove (5.3.17). Let $v_h\to v$ weakly in $L^1(0,T;\mathbf{R}^n)$, $y_h\to y$ in $L^\infty(0,T;\mathbf{R}^n)$ and assume that for infinitely many $h\in\mathbf{N}$ we have

$$\begin{cases} y_h' = A(t,y_h) + B(t,y_h)\,v_h & \text{on } [0,T] \\ y_h(0)\in K. \end{cases}$$

Since K is closed we have $y(0)\in K$; moreover, setting

(5.3.19) $\quad g_h(t,y) = A(t,y) + B(t,y)\, v_h(t) \qquad (t\in[0,T],\ y\in\mathbf{R}^n)$

and applying Proposition 5.3.3 and Remark 5.3.4 we obtain

$$y' = A(t,y) + B(t,y)\, v(t) \quad \text{on } [0,T].$$

Let us prove (5.3.18). Assume that

$$\begin{cases} y' = A(t,y) + B(t,y)\, v & \text{on } [0,T] \\ y(0)\in K \end{cases}$$

and let $v_h \to v$ weakly in $L^1(0,T;\mathbf{R}^k)$. Setting $g_h(t,y)$ as in (5.3.19) and applying Proposition 5.3.3 and Remark 5.3.4 we get that for h large enough there exists $y_h\in W^{1,1}(0,T;\mathbf{R}^n)$ such that

$$\begin{cases} y_h' = A(t,y_h) + B(t,y_h)\, v_h & \text{on } [0,T] \\ y_h(0) = y(0) \end{cases}$$

and $y_h \to y$ in $L^\infty(0,T;\mathbf{R}^n)$, and this achieves the proof. ■

Recalling Proposition 5.2.1, Corollary 5.1.5, Lemma 5.3.2, and Proposition 5.3.4 we obtain the following form for the relaxed functional $\Gamma F = \Gamma_{seq}(U^-,Y^-)F$:

$$\Gamma F(u,y) = \inf\Big\{ \int_0^T \varphi^{**}(t,y,u,v)\, dt : v\in V,\ y' = A(t,y)+B(t,y)v \text{ on } [0,T],\ y(0)\in K \Big\}$$

where $\varphi(t,y,u,v) = f(t,y,u) + \chi_{\{v=b(t,u)\}}$ and the convexification is made with respect to the variables (u,v). Now, we eliminate the variable v in the last expression. Denote by $\underline{co}$ the convex closed hull operator, and set

$$\beta(t,u) = \{v\in\mathbf{R}^k : (u,v)\in \underline{co}\{(\lambda,\mu)\in\mathbf{R}^m\times\mathbf{R}^k : \mu = b(t,\lambda)\}\},$$

$$\bar{f}(t,y,u,w) = \inf\big\{\varphi^{**}(t,y,u,v) : w = A(t,y)+B(t,y)\, v\big\},$$

$$\bar{\Lambda} = \big\{(u,y)\in U\times Y : y'\in A(t,y)+B(t,y)\,\beta(t,u) \text{ on } [0,T],\ y(0)\in K\big\}.$$

The following result holds.

THEOREM 5.3.6. *Under the assumptions (5.3.3), (5.3.4), (5.3.5), (5.3.15), (5.3.16) the relaxed functional* ΓF *can be written in the form*

$$\Gamma F(u,y) = \int_0^T \bar{f}(t,y,u,y')\, dt + \chi_{\bar{\Lambda}}(u,y) .$$

Proof. It is easy to see that $v \in \beta(t,u)$ whenever $\varphi^{**}(t,y,u,v) < +\infty$; hence

$$\Gamma F(u,y) = \chi_{\bar{\Lambda}}(u,y) + \inf\Big\{ \int_0^T \varphi^{**}(t,y,u,v)\, dt : v \in V,\ \ y' = A(t,y) + B(t,y)\, v \ \text{ on } [0,T] \Big\} .$$

Moreover, for every $v \in V$ such that $y' = A(t,y) + B(t,y)\, v$ on $[0,T]$, we have

$$\varphi^{**}(t,y(t),u(t),v(t)) \geq \bar{f}(t,y(t),u(t),y'(t)) ,$$

so that

$$\Gamma F(u,y) = \chi_{\bar{\Lambda}}(u,y) + \int_0^T \bar{f}(t,y,u,y')\, dt .$$

In order to prove the opposite inequality, fix $(u,y) \in U \times Y$ and assume

$$(u,y) \in \bar{\Lambda} \qquad \text{and} \qquad \int_0^T \bar{f}(t,y,u,y')\, dt < +\infty .$$

Then $f(t,y(t),u(t),y'(t)) < +\infty$ a.e. on $[0,T]$, and by definition of f and hypotheses (5.3.3), (5.3.4), (5.3.5) we have that for a.e. $t \in [0,T]$

$$\bar{f}(t,y(t),u(t),y'(t)) = \varphi^{**}(t,y(t),u(t),y'(t))$$

for a suitable $v(t) \in \mathbf{R}^k$ such that $y'(t) = A(t,y(t)) + B(t,y(t))\, v(t)$. By Proposition 2.2.7 the function $v(t)$ can be choosen measurable, and by (5.3.3), (5.3.4), (5.3.5) we have $v \in L^1(0,T;\mathbf{R}^k)$. Therefore

$$\Gamma F(u,y) \le \chi_{\overline{\Lambda}}(u,y) + \int_0^T \varphi^{**}(t,y,u,v)\,dt =$$

$$= \chi_{\overline{\Lambda}}(u,y) + \int_0^T \overline{f}(t,y,u,y')\,dt . \blacksquare$$

EXAMPLE 5.3.7. Let $n=m=1$; consider the following optimal control problem:

minimize the cost functional $$\int_0^1 \Big[u^2 + \frac{1}{u^2} + |y - y_0(t)|^2 + h(t)\,u\Big]\,dt$$

under the constraints $uy'=1$, $u>0$, $y(0)\in K$.

Here $y_0(t)$ and $h(t)$ are two functions in $L^2(0,1)$, and K is a closed subset of $\mathbf{R}$. In order to compute the relaxed problem we remark that, due to the growth of the cost functional, the relaxation does not change if we consider on the space of controls the weak $L^2(0,1)$ topology. Moreover, since the term

$$\int_0^1 \Big[|y - y_0(t)|^2 + h(t)\,u\Big]\,dt$$

is continuous with respect to the convergence $L^\infty\times(\text{weak } L^2)$, by Proposition 5.1.2 we may reduce ourselves to relax the problem

$$\int_0^1 \Big(u^2 + \frac{1}{u^2}\Big)\,dt + \chi_{\{uy'=1,\ u>0,\ y(0)\in K\}} .$$

With the notation of Theorem 5.3.5 it is

$$A(t,y) \equiv 0, \quad B(t,y) \equiv 1, \quad b(t,u) = \frac{1}{u},$$

$$f(u) = u^2 + \frac{1}{u^2} + \chi_{\{u>0\}}, \quad \varphi(u,v) = u^2 + v^2 + \chi_{\{uv=1,\ u>0\}}$$

and it is immediately seen that conditions (5.3.3), (5.3.4), (5.3.5), (5.3.15), (5.3.16) are fulfilled. The multimapping $\beta(t,u)$ is given by

$$b(t,u) = \{v \in \mathbf{R} : uv \geq 1\},$$

so it remains only to compute the function $\bar{f}(t,y,u,w)$ which coincides, in this case, with $\varphi^{**}(u,w)$. Denote by $g(u,w)$ the convex l.s.c. function

$$g(u,w) = u^2 + w^2 + 2(uw - 1) + \chi_{\{uw \geq 1,\, u>0\}} ;$$

since it is $g \leq \varphi$ it is also

$$g(u,w) \leq \varphi^{**}(u,w) .$$

On the other hand, if $u>0$ and $uw \geq 1$, let $a,b,\alpha,\beta,\lambda,\mu$ be the positive solutions of

$$\begin{cases} \lambda + \mu = 1 \\ a\,b = 1 \\ \alpha\,\beta = 1 \\ a + b = \alpha + \beta \\ (u,w) = \lambda\,(a,b) + \mu\,(\alpha,\beta) ; \end{cases}$$

we have

$$\begin{aligned} \varphi^{**}(u,w) &\leq \lambda\,\varphi(a,b) + \mu\,\varphi(a,b) = \lambda\,(a^2+b^2) + \mu\,(\alpha^2+\beta^2) = \\ &= \lambda\,[(a+b)^2 - 2ab] + \mu\,[(\alpha+\beta)^2 - 2\alpha\beta] = \\ &= (a+b)^2 - 2 = a^2+b^2 . \end{aligned}$$

But $u+w=a+b$; therefore, from $ab=1$ it follows that $a^2-a(u+w)+1 = 0$, and so

$$\begin{aligned} \varphi^{**}(u,w) &\leq a^2 + (u+w-a)^2 = 2a^2 + (u+w)^2 - 2a(u+w) = \\ &= (u+w)^2 - 2 = u^2 + w^2 + 2(uw-1) = g(u,w). \end{aligned}$$

Then, the relaxed control problem is

$$\min \left\{ \int_0^1 \left[u^2 + y'^2 + 2(uy'-1) + |y - y_0(t)|^2 + h(t)\,u\right] dt \ : \ uy' \geq 1,\, u>0,\, y(0) \in K \right\} .$$

EXAMPLE 5.3.8. Consider the same problem as in Example 5.3.7, but with the additional constraint on the control:

$$\frac{1}{c} \leq u \leq c \qquad \text{for a suitable } c \geq 1.$$

By the same kind of calculations of Example 5.3.7 one finds that the relaxed problem is the minimization problem for

$$\int_0^1 \left[u^2 + y'^2 + 2(uy'-1) + |y - y_0(t)|^2 + h(t)\, u\right] dt$$

with the constraints

$$\frac{1}{u} \leq y' \leq c + \frac{1}{c} - u, \quad \frac{1}{c} \leq u \leq c, \quad y(0) \in K.$$

5.4. Problems with Elliptic State Equations

In this section we consider optimal control problems governed by elliptic partial differential equations. Let m,n be positive integers and let Ω be a bounded open subset of $\mathbf{R}^n$ with a Lipschitz boundary. We denote by Y the space $H^1(\Omega)$ endowed with its weak topology, and by U the space $L^1(\Omega;\mathbf{R}^m)$ endowed with its weak topology. The cost functional we consider is of the form

$$J(u,y) = \int_\Omega f(x,y,u)\, dx$$

where $f:\Omega\times\mathbf{R}\times\mathbf{R}^m \to [0,+\infty]$ is a given Borel function. In order to define the admissible set Λ, we consider an elliptic operator

$$A = -\sum_{i,j=1}^{n} D_i\left(a_{ij}(x)\, D_j\right)$$

with $a_{ij}=a_{ji}\in L^{\infty}(\Omega)$ and

$$M_0\,|z|^2 \le \sum_{i,j=1}^{n} a_{ij}(x)\,z_i\,z_j \le M_1\,|z|^2 \qquad \forall(x,z)\in\Omega\times\mathbf{R}^n \qquad (0<M_0\le M_1).$$

The admissible set Λ is then defined by

$$\Lambda = \{(u,y)\in U\times Y \;:\; Ay=b(x,u) \text{ on } \Omega\}$$

where $b:\Omega\times\mathbf{R}^m\to\mathbf{R}$ is a given Borel function. Finally, the functional to be relaxed is

$$F(u,y) = J(u,y)+\chi_\Lambda(u,y).$$

About the function f we assume that

(5.4.1) there exists $\alpha\in[1,+\infty]$ such that for every $x\in\Omega$, $u\in\mathbf{R}^m$, $y,z\in\mathbf{R}$

$$f(x,y,u) \le f(x,z,u)+\rho(x,|y-z|)+\sigma(x,|y-z|)\,[f(x,z,u)]^{(\alpha-1)/\alpha}$$

where $\rho(x,s)$ and $\sigma(x,s)$ are two functions from $\Omega\times[0,+\infty[$ into $[0,+\infty[$ measurable in x, continuous and increasing in s, with $\rho(x,0)=\sigma(x,0)=0$, and such that the operators $z\to\rho(x,|z(x)|)$ and $z\to\sigma(x,|z(x)|)$ are sequentially continuous from Y into $L^1(\Omega)$ and $L^{\alpha}(\Omega)$ respectively;

(5.4.2) there exist $a\in L^1(\Omega)$, $c>0$, and a function $\psi:\mathbf{R}\to\mathbf{R}$ (which can be taken increasing and convex) such that

$$\lim_{s\to+\infty}\frac{\psi(s)}{s} = +\infty$$

$$\psi(|u|)+c\,|b(x,u)|^{2n/(n+2)}-a(x) \le f(x,0,u) \qquad \forall(x,u)\in\Omega\times\mathbf{R}^m;$$

(5.4.3) there exists $u_0\in L^1(\Omega;\mathbf{R}^m)$ such that the function $f(x,0,u_0(x))$ is integrable on Ω.

REMARK 5.4.1. As usual, the inequality in (5.4.2) becomes

$$\psi(|u|)+c|b(x,u)|^p-a(x) \le f(x,0,u) \qquad \forall(x,u)\in\Omega\times\mathbf{R}^m$$

for some $p>1$ if $n=2$, and

$$\psi(|u|+|b(x,u)|) - a(x) \le f(x,0,u) \qquad \forall (x,u)\in \Omega\times\mathbf{R}^m$$

if $n=1$.

As in the previous section, we apply the "auxiliary variable" method to the functional F. In the following we consider only the case $n>2$ and we denote by p the number ${}^{2n}/_{(n+2)}$; the obvious modifications in the other cases can be made by taking into account Remark 5.4.1.

LEMMA 5.4.2. *Let* V *be the space* $L^p(\Omega)$ *endowed with its weak topology, and let* $\vartheta: U\times Y \to V$ *be given by*

$$[\vartheta(u,y)](x) = \begin{cases} b(x,u(x)) & \textit{if } b(x,u(x))\in L^p(\Omega) \\ 0 & \textit{otherwise.} \end{cases}$$

Then, the compactness property (5.2.1) holds for the functional F.

Proof. Let (u_h,y_h) be a converging sequence in $U\times Y$ such that

(5.4.4) $\qquad F(u_h,y_h) \le C \qquad \forall h\in\mathbf{N},$

and let $v_h(x)=b(x,u_h(x))$. By (5.4.1), (5.4.2), (5.4.4) we get

$$\int_\Omega \big[c\,|v_h|^p - a(x)\big]\,dx \le \int_\Omega f(x,0,u_h)\,dx \le$$

$$\le \int_\Omega \big[f(x,y_h,u_h) + \rho(x,|y_h|)\big]\,dx + \|\sigma(x,|y_h|)\|_{L^\alpha(\Omega)} \Big(\int_\Omega f(x,y_h,u_h)\,dx\Big)^{(\alpha-1)/\alpha} \le$$

$$\le C + \|\rho(x,|y_h|)\|_{L^1(\Omega)} + C^{(\alpha-1)/\alpha}\|\sigma(x,|y_h|)\|_{L^\alpha(\Omega)}$$

which is bounded, by the hypotheses on ρ and σ. Then $\{v_h\}$ is compact in V. ■

By Proposition 5.2.1 the study of the relaxation of the functional F is reduced to the relaxation in $(U\times V)\times Y$ of the functional

$$G(u,v,y) = \Phi(u,v,y) + \chi_{\Theta}(u,v,y)$$

where

$$\Phi(u,v,y) = \int_{\Omega} \varphi(x,y,u,v)\,dx$$

$$\varphi(x,y,u,v) = f(x,y,u) + \chi_{\{v=b(x,u)\}}$$

$$\Theta = \{(u,v,y)\in U\times V\times Y \ : \ Ay=v\}.$$

As in the previous section, we compute separately

$$\Gamma_{seq}((U\times V)^{-},Y)\Phi \quad \text{and} \quad \Gamma_{seq}(U\times V,Y^{-})\chi_{\Theta}\,.$$

LEMMA 5.4.3. *For every* $(u,v,y)\in U\times V\times Y$ *we have*

$$\Gamma_{seq}((U\times Y)^{-},Y)\Phi(u,v,y) = \int_{\Omega} \varphi^{**}(x,y,u,v)\,dx\,.$$

Proof. By property (5.4.1) we obtain

$$\Phi(u,v,y) \le \Phi(u,v,z) + \|\rho(x,|y-z|)\|_{L^1(\Omega)} + \|\sigma(x,|y-z|)\|_{L^{\alpha}(\Omega)} \big[\Phi(u,v,z)\big]^{(\alpha-1)/\alpha}$$

whenever $u\in U$, $v\in V$, $y,z\in Y$. Hence, conditions (5.2.3), (5.2.4), (5.2.5) of Proposition 5.2.2 are satisfied, with

$$\omega(y,z) = \|\rho(x,|y-z|)\|_{L^1(\Omega)} + \|\sigma(x,|y-z|)\|_{L^{\alpha}(\Omega)}$$

$$H(u,v,z) = 1 + \big[\Phi(u,v,z)\big]^{(\alpha-1)/\alpha},$$

and so, for every $(u,v,y)\in U\times V\times Y$ we have

$$\Gamma_{seq}((U\times V)^{-},Y)\Phi(\upsilon,\varpi,\psi) = \Gamma_{seq}((U\times V)^{-},\delta_Y)\Phi(u,v,y).$$

Therefore, the same argument used in the proof of Lemma 5.3.2 gives

$$\Gamma_{seq}((U\times Y)^{-},Y)\Phi(u,v,y) = \int_{\Omega} \varphi^{**}(x,y,u,v)\, dx \, . \blacksquare$$

LEMMA 5.4.4. *For every* $(u,v,y)\in U\times V\times Y$ *we have*

$$\Gamma_{seq}(U\times V,Y^{-})\chi_{\Theta}(u,v,y) = \chi_{\Theta}(u,v,y).$$

Proof. Fix $(u,v,y)\in U\times V\times Y$; we have to prove that

(i) $(v_h,y_h)\to(v,y)$ in $V\times Y$, $Ay_h=v_h$ for infinitely many $h\in\mathbf{N}$ $\Rightarrow$ $Ay=v$;

(ii) $Ay=v$ $\Rightarrow$ $\forall v_h\to v$ $\exists y_h\to y$: $Ay_h=v_h$ for every $h\in\mathbf{N}$ large enough.

Since A is a linear elliptic operator and $^{2n}/_{(n+2)}=(2^*)'$, (i) and (ii) follow straightforward from the Sobolev imbedding theorem. ■

We may conclude the study of the relaxation of F by the same argument used in the previous section, and we obtain

$$\Gamma F(u,y) = \int_{\Omega} \varphi^{**}(x,y,u,Ay)\, dx + \chi_{\overline{\Lambda}}(u,y)$$

where

$$\overline{\Lambda} = \{(u,y)\in U\times Y : Ay\in\beta(x,u) \text{ on } \Omega\}$$

$$\beta(x,u) = \{v\in\mathbf{R} : (u,v)\in\underline{co}\{(\lambda,\mu)\in\mathbf{R}^m\times\mathbf{R} : \mu=b(x,\lambda)\}\}.$$

EXAMPLE 5.4.5. With the same computations made in Examples 5.3.7 and 5.3.8 we may show that for every $y_0\in L^2(\Omega)$ and $h\in L^2(\Omega)$ the control problem with cost functional

$$\int_\Omega \left[u^2 + \frac{1}{u^2} + |y - y_0(x)|^2 + h(x)\,u\right] dx ,$$

control constraints $1/_c \leq u \leq c$ for some $c \geq 1$, and state equation

$$\begin{cases} u\,Ay = 1 & \text{in } \Omega \\ y \in H_0^1(\Omega) , \end{cases}$$

relaxes to the problem

$$\min \left\{ \int_\Omega \left[u^2 + |Ay|^2 + 2(u\,Ay - 1) + |y - y_0(x)|^2 + h(x)\,u\right] dx \right\}$$

subject to $1/_c \leq u \leq c$ and

$$\begin{cases} \frac{1}{u} \leq Ay \leq c + \frac{1}{c} - u & \text{in } \Omega \\ y \in H_0^1(\Omega) . \end{cases}$$

Bibliography

[1] F.ACANFORA & M.SQUILLANTE: *Sulla semicontinuità inferiore degli integrali multipli di ordine superiore*. Ricerche Mat., **30** (1981), 333-354.

[2] E.ACERBI & G.BUTTAZZO: *Semicontinuous envelopes of polyconvex integrals*. Proc.Roy.Soc.Edinburgh, **96-A** (1984), 51-54.

[3] E.ACERBI & G.BUTTAZZO & N.FUSCO: *Semicontinuity in L^∞ for polyconvex integrals*. Atti Accad.Naz.Lincei Rend.Cl.Sci.Fis.Mat.Natur., **72** (1982), 25-28.

[4] E.ACERBI & G.BUTTAZZO & N.FUSCO: *Semicontinuity and relaxation for integrals depending on vector-valued functions*. J.Math.Pures Appl., **62** (1983), 371-387.

[5] E.ACERBI & N.FUSCO: *Semicontinuity problems in the calculus of variations*. Arch.Rational Mech.Anal., **86** (1984), 125-145.

[6] R.A.ADAMS: *Sobolev Spaces*. Academic Press, New York (1975).

[7] N.I.AKHIEZER: *The Calculus of Variations*. Blaisdell, New York (1962).

[8] A.A.ALBERT: *A quadratic form problem in the calculus of variations*. Bull. Amer.Math.Soc., **44** (1938), 250-253.

[9] L.AMBROSIO: *Problemi di semicontinuità e rilassamento in calcolo delle variazioni*. Tesi di Laurea, University of Pisa, Pisa (1985).

[10] L.AMBROSIO: *Nuovi risultati sulla semicontinuità inferiore di certi funzionali integrali*. Atti Accad.Naz.Lincei Rend.Cl.Sci.Fis.Mat.Natur., **79** (1985), 82-89.

[11] L.AMBROSIO: *New lower semicontinuity results for integral functionals*. Rend.Accad.Naz.Sci.XL Mem.Mat.Sci.Fis.Natur., **11** (1987), 1-42.

[12] L.AMBROSIO: *Relaxation of autonomous functionals with discontinuous integrands*. Ann.Univ.Ferrara, (to appear).

[13] L.AMBROSIO: *A lower closure theorem for autonomous orientor fields*. Proc. Roy.Soc.Edinburgh, (to appear).

[14] L.AMBROSIO & G.BUTTAZZO: *Weak lower semicontinuous envelope of functionals defined on a space of measures*. Ann.Mat.Pura Appl., **150** (1988), 311-340.

[15] L.AMBROSIO & G.BUTTAZZO & A.LEACI: *Continuous operators of the form $T_f(u)=f(x,u,Du)$*. Boll.Un.Mat.Ital., (to appear).

[16] G.ANZELLOTTI: *The Euler equation for functionals with linear growth*. Trans.Amer.Math.Soc., **290** (1985), 483-501.

[17] G.ANZELLOTTI & G.BUTTAZZO & G.DAL MASO: *Dirichlet problems for demi-coercive functionals*. Nonlinear Anal., **10** (1986), 603-613.

[18] G.ANZELLOTTI & M.GIAQUINTA: *Funzioni BV e tracce*. Rend.Sem.Mat. Univ.Padova, **60** (1978), 1-22.

[19] J.APPELL: *The Superposition Operator in Function Spaces. A Survey*. Book in preparation.

[20] J.APPELL & P.P.ZABREJKO: *Continuity properties of the superposition operator*. Preprint University of Augsburg, Augsburg (1987).

[21] H.ATTOUCH: *Variational Convergence for Functions and Operators*. Appl. Math.Ser., Pitman, Boston (1984).

[22] H.ATTOUCH & C.PICARD: *Problèmes variationnels et théorie du potentiel non linéaire*. Ann.Fac.Sci.Toulouse Math., **1** (1979), 89-136.

[23] H.ATTOUCH & R.WETS: *Isometries for the Legendre-Fenchel transform*. Trans.Amer.Math.Soc., **296** (1986), 33-60.

[24] G.AUBERT & R.TAHRAOUI: *Théorèmes d'existence pour des problèmes du calcul des variations*. J.Differential Equations, **33** (1979), 1-15.

[25] G.AUBERT & R.TAHRAOUI: *Sur la minimisation d'une fonctionnelle non convexe, non différentiable en dimension 1*. Boll.Un.Mat.Ital., **17-B** (1980), 244-258.

[26] C.BAIOCCHI & G.BUTTAZZO & F.GASTALDI & F.TOMARELLI: *General existence results for unilateral problems in continuum mechanics*. Arch. Rational Mech.Anal., **100** (1988), 149-189.

[27] E.J.BALDER: *Lower closure problems with weak convergence conditions in a new perspective*. SIAM J.Control Optim., **20** (1982), 198-210.

[28] E.J.BALDER: *A general approach to lower semicontinuity and lower closure in optimal control theory*. SIAM J.Control Optim., **4** (1984), 570-598.

[29] E.J.BALDER: *Lower closure for orientor fields by lower semicontinuity of outer integral functionals*. Ann.Mat.Pura Appl., **139** (1985), 349-359.

[30] E.J.BALDER: *On seminormality of integral functionals and their integrands*. SIAM J.Control Optim., **24** (1986), 95-121.

[31] E.J.BALDER: *Necessary and sufficient conditions for L_1-strong-weak lower semicontinuity of integral functionals*. Nonlinear Anal., **11** (1987), 1399-1404.

[32] J.M.BALL: *On the calculus of variations and sequentially weakly continuous maps*. Proceedings of "Ordinary and Partial Differential Equations", Dundee 1976, edited by W.N.Everitt & B.D.Sleeman, Lecture Notes in Math. **564**, Springer-Verlag, Berlin (1976), 13-23.

[33] J.M.BALL: *Convexity conditions and existence theorems in nonlinear elasticity*. Arch.Rational Mech.Anal., **63** (1977), 337-403.

[34] J.M.BALL: *Constitutive inequalities and existence theorems in nonlinear elasticity*. Proceedings of "Nonlinear Analysis and Mechanics", Edinburgh 1976, edited by R.J.Knops, Res.Notes in Math. **17**, Pitman, Boston (1977), 13-25.

[35] J.M.BALL: *Strict convexity, strong ellipticity, and regularity in the calculus of variations*. Math.Proc.Cambridge Philos.Soc., **87** (1980), 501-513.

[36] J.M.BALL: *Remarks on the paper "Basic calculus of variations"*. Pacific J. Math., **116** (1985), 7-10.

[37] J.M.BALL: *Does rank one convexity imply quasiconvexity?* Proceedings of "Metastability and Incompletely Posed Problems", Minneapolis 1985, edited by S.S.Antman & J.L.Ericksen & D.Kinderlehrer & I.Müller, Springer-Verlag, Berlin (1987), 17-32.

[38] J.M.BALL & J.C.CURRIE & P.J.OLVER: *Null lagrangians, weak continuity, and variational problems of arbitrary order*. J.Funct.Anal., **41** (1981), 135-174.

[39] J.M.BALL & J.E.MARSDEN: *Quasiconvexity at the boundary, positivity of the second variation, and elastic stability*. Arch.Rational Mech.Anal., **86** (1984), 251-277.

[40] J.M.BALL & V.J.MIZEL: *One dimensional Variational Problems whose minimizers do not satisfy the Euler-Lagrange Equation*. Arch.Rational.Mech.Anal. **90** (1985), 325-388.

[41] J.M.BALL & V.J.MIZEL: *Singular minimizers for regular one-dimensional problems in the calculus of variations*. Bull.Amer.Math.Soc., **11** (1985), 143-146.

[42] J.M.BALL & F.MURAT: *$W^{1,q}$ quasiconvexity and variational problems for multiple integrals*. J.Funct.Anal., **58** (1984), 225-253.

[43] C.BENASSI & A.GAVIOLI: *Some results about relaxation of integral functionals*. Atti Sem.Mat.Fis.Univ.Modena, **35** (1987), 289-307.

[44] A.BENSOUSSAN & J.L.LIONS & G.PAPANICOLAOU: *Asymptotic Analysis for Periodic Structures*. North-Holland, Amsterdam (1978).

[45] H.BERGSTROM: *Weak Convergence of Measures*. Academic Press, New York (1982).

[46] L.D.BERKOWITZ: *Optimal Control Theory*. Springer-Verlag, Berlin (1974).

[47] L.D.BERKOWITZ: *Lower semicontinuity of integral functionals*. Trans.Amer. Math.Soc., **192** (1974), 51-57.

[48] H.BERLIOCCHI & J.M.LASRY: *Intégrandes normales et mesures paramétrées en calcul des variations*. Bull.Soc.Math.France, **101** (1973), 129-184.

[49] N.BERRUTI ONESTI: *Sopra la semicontinuità di una classe di integrali curvilinei per problemi variazionali di ordine n*. Istit.Lombardo Accad.Sci.Lett. Rend., **A-106** (1972), 365-396.

[50] G.A.BLISS: *Lectures on the Calculus of Variations*. University of Chicago Press, Chicago (1946).

[51] L.BOCCARDO & G.BUTTAZZO: *Quasilinear elliptic equations with discontinuous coefficients*. Atti Accad. Naz.Lincei Rend.Cl.Sci.Fis.Mat.Natur., (to appear).

[52] O.BOLZA: *Lectures on the Calculus of Variations*. Chelsea Publishing Company, New York (1946).

[53] M.BONI: *Un teorema di semicontinuità inferiore*. Rend.Circ.Mat.Palermo, **25** (1976), 53-66.

[54] M.BORGOGNO: *Sopra le condizioni necessarie per la semicontinuità degli integrali dei problemi variazionali in forma parametrica di secondo ordine*. Istit. Lombardo Accad.Sci.Lett.Rend., **A-108** (1974), 228-261.

[55] G.BOTTARO & P.OPPEZZI: *Condizioni necessarie per la semicontinuità inferiore di un funzionale integrale dipendente da funzioni a valori in uno spazio di Banach*. Boll.Un.Mat.Ital., **18-B** (1981), 47-65.

[56] G.BOTTARO & P.OPPEZZI: *Rappresentazione con integrali multipli di funzionali dipendenti da funzioni a valori in uno spazio di Banach*. Ann.Mat.Pura Appl., **139** (1985), 191-225.

[57] G.BOUCHITTE: *Représentation intégrale de fonctionnelles convexes sur un espace de mesures*. Ann.Univ.Ferrara, **33** (1987), 113-156.

[58] G.BOUCHITTE & M.VALADIER: *Integral representation of convex functionals on a space of measures*. Preprint University of Montpellier, Montpellier (1988).

[59] P.BRANDI: *Teoremi di semicontinuità inferiore e di approssimazione per le variazioni con peso in* $\tilde{\mathbb{R}}$. Rend.Circ.Mat.Palermo, **25** (1976), 5-26.

[60] P.BRANDI & A.SALVADORI: *Existence, semicontinuity and representation for the integrals of the calculus of variations. The BV case*. Rend.Circ.Mat. Palermo, **8** (1985), 447-462.

[61] H.BREZIS: *Intégrales convexes dans les espaces de Sobolev*. Israel J.Math., **13** (1972), 9-23.

[62] H.BREZIS: *Opérateurs Maximaux Monotones*. North-Holland, Amsterdam (1973).

[63] H.BREZIS: *Analyse Fonctionnelle et Applications*. Masson, Paris (1983).

[64] F.E.BROWDER: *Remarks on the direct method of the calculus of variations*. Arch.Rational Mech.Anal., **20** (1965), 251-258.

[65] G.BUTTAZZO: *Su una definizione generale dei Γ-limiti*. Boll.Un.Mat.Ital., **14-B** (1977), 722-744.

[66] G.BUTTAZZO: *Problemi di semicontinuità e rilassamento in calcolo delle variazioni*. Proceedings of "Equazioni Differenziali e Calcolo delle Variazioni", Pisa 1985, edited by L.Modica, ETS Editrice, Pisa (1985), 23-36.

[67] G.BUTTAZZO: *Semicontinuity, relaxation, and integral representation problems in the calculus of variations*. Notes of a series of lectures held at CMAF of Lisbon in November-December 1985. Printed by CMAF, Lisbon (1986).

[68] G.BUTTAZZO: *Some relaxation problems in optimal control theory*. J.Math. Anal.Appl., **125** (1987), 272-287.

[69] G.BUTTAZZO: *Relaxation problems in control theory*. Proceedings of "Calculus of Variations and Partial Differential Equations", Trento 1986, edited by S. Hildebrandt & D.Kinderlehrer & M.Miranda, Lecture Notes in Math. **1340**, Springer-Verlag, Berlin (1988), 31-39.

[70] G.BUTTAZZO & G.DAL MASO: *Integral representation on $W^{1,\alpha}(\Omega)$ and $BV(\Omega)$ of limits of variational integrals*. Atti Accad.Naz.Lincei Rend.Cl.Sci. Fis.Mat.Natur., **66** (1979), 338-343.

[71] G.BUTTAZZO & G.DAL MASO: *Γ-limits of integral functionals*. J.Analyse Math., **37** (1980), 145-185.

[72] G.BUTTAZZO & G.DAL MASO: *Γ-convergence and optimal control problems*. J.Optim.Theory Appl., **38** (1982), 385-407.

[73] G.BUTTAZZO & G.DAL MASO: *On Nemyckii operators and integral representation of local functionals*. Rend.Mat., **3** (1983), 491-509.

[74] G.BUTTAZZO & G.DAL MASO: *A characterization of nonlinear functionals on Sobolev spaces which admit an integral representation with a Carathéodory integrand*. J.Math.Pures Appl., **64** (1985), 337-361.

[75] G.BUTTAZZO & G.DAL MASO: *Integral representation and relaxation of local functionals*. Nonlinear Anal., **9** (1985), 512-532.

[76] G.BUTTAZZO & G.DAL MASO & U.MOSCO: *A derivation theorem for capacities with respect to a Radon measure*. J.Funct.Anal., **71** (1987), 263-278.

[77] G.BUTTAZZO & A.LEACI: *A continuity theorem for operators from $W^{1,q}(\Omega)$ into $L^r(\Omega)$*. J.Funct.Anal., **58** (1984), 216-224.

[78] G.BUTTAZZO & A.LEACI: *Relaxation results for a class of variational integrals*. J.Funct.Anal., **61** (1985), 360-377.

[79] O.CALIGARIS & F.FERRO & P.OLIVA: *Sull'esistenza del minimo per problemi di calcolo delle variazioni relativi ad archi di variazione limitata*. Boll.Un. Mat.Ital., **14-B** (1977), 340-349.

[80] O.CALIGARIS & P.OLIVA: *Sulla caratterizzazione di problemi di calcolo delle variazioni per funzioni di variazione limitata*. Boll.Un.Mat.Ital., **15-B** (1978), 253-271.

[81] C.CARATHEODORY: *Calculus of Variations and Partial Differential Equations of the First Order. I.II*. Holden Day, San Francisco (1965) and (1967).

[82] L.CARBONE & C.SBORDONE: *Some properties of Γ-limits of integral functionals*. Ann.Mat.Pura Appl., **122** (1979), 1-60.

[83] M.CARRIERO & G.DAL MASO & A.LEACI & E.PASCALI: *Relaxation of the non-parametric Plateau problem with an obstacle*. J.Math.Pures Appl., (to appear).

[84] M.CARRIERO & A.LEACI & E.PASCALI: *Integrals with respect to a Radon measure added to area type functionals: semicontinuity and relaxation*. Atti Accad.Naz.Lincei Rend.Cl.Sci.Fis.Mat.Natur., **78** (1985), 133-137.

[85] M.CARRIERO & A.LEACI & E.PASCALI: *Semicontinuità e rilassamento per funzionali somma di integrali del tipo dell'area e di integrali rispetto ad una misura di Radon*. Rend.Accad.Naz.Sci.XL Mem.Mat.Sci.Fis.Natur., **10** (1986), 1-31.

[86] M.CARRIERO & A.LEACI & E.PASCALI: *On the semicontinuity and the relaxation for integrals with respect to the Lebesgue measure added to integrals with respect to a Radon measure*. Ann.Mat.Pura Appl., **149** (1987), 1-21.

[87] C.CASTAING & P.CLAUZURE: *Semicontinuité des fonctionnelles intégrales*. Acta Math.Vietnam., **7** (1984), 139-170.

[88] C.CASTAING & M.VALADIER: *Convex Analysis and Measurable Multifunctions*. Lecture Notes in Math. **580**, Springer-Verlag, Berlin (1977).

[89] L.CESARI: *Semicontinuità e convessità nel calcolo delle variazioni*. Ann. Scuola Norm.Sup.Pisa Cl.Sci., **18** (1964), 389-423.

[90] L.CESARI: *Lower semicontinuity and lower closure theorems without seminormality conditions*. Ann.Mat.Pura Appl., **98** (1974), 381-397.

[91] L.CESARI: *A necessary and sufficient condition for lower semicontinuity*. Bull.Amer.Math.Soc., **80** (1974), 467-472.

[92] L.CESARI: *Optimization-Theory and Applications*. Springer-Verlag, New York (1983).

[93] L.CESARI: *Nonlinear analysis*. Boll.Un.Mat.Ital., **A-4** (1985), 157-216.

[94] L.CESARI & P.BRANDI & A.SALVADORI: *Existence theorems concerning simple integrals of the calculus of variations for discontinuous solutions*. Arch. Rational Mech.Anal., **98** (1987), 307-328.

[95] S.CINQUINI: *Recent results about the semicontinuity of a class of integrals in the calculus of variations*. Rend.Sem.Mat.Fis.Milano, **52** (1982), 497-516.

[96] S.CINQUINI: *Further investigation of the semicontinuity of a class of integrals in the calculus of variations*. Atti Accad.Sci.Istit.Bologna Cl.Sci.Fis.Rend., **10** (1984), 25-42.

[97] F.H.CLARKE: *The Euler-Lagrange differential inclusion*. J.Differential Equations, **19** (1975), 80-90.

[98] F.H.CLARKE: *Admissible relaxation in variational and control problems*. J. Math.Anal.Appl., **51** (1975), 557-576.

[99] F.H.CLARKE: *The Erdmann condition and Hamiltonian inclusions in optimal control and the calculus of variations*. Canad.J.Math., **32** (1980), 494-509.

[100] F.H.CLARKE: *Optimization and Nonsmooth Analysis*. Wiley Interscience, New York (1983).

[101] F.H.CLARKE: *Tonelli's regulerity theory in the calculus of variations: recent progress*. Proceedings of "Optimization and Related Fields", Erice 1984, edited by R.Conti & E.De Giorgi & F.Giannessi, Lecture Notes in Math. **1190**, Springer-Verlag, Berlin (1986), 163-179.

[102] F.H.CLARKE & R.B.VINTER: *On the conditions under which the Euler equation or the maximum principle hold*. Appl.Math.Optim., **12** (1984), 73-79.

[103] F.H.CLARKE & R.B.VINTER: *Regularity properties of solutions to the basic problem in the calculus of variations*. Trans.Amer.Math.Soc., **289** (1985), 73-98.

[104] F.H.CLARKE & R.B.VINTER: *Existence and regularity in the small in the calculus of variations*. J.Differential Equations, **59** (1985), 336-354.

[105] F.COLOMBINI & E.DE GIORGI & L.C.PICCININI: *Frontiere Orientate di Misura Minima e Questioni Collegate*. Quaderno della Scuola Normale Superiore, Pisa (1972).

[106] M.CORAL: *On a necessary condition for the minimum of a double integral*. Duke Math.J., **3** (1937), 585-592.

[107] C.COSTANTINESCU & K.WEBER & A.SONTAG: *Integration Theory. Volume 1: Measure and Integral*. Wiley, New York (1985).

[108] R.COURANT: *Calculus of Variations*. Courant Institute Publications, New York (1962).

[109] B.DACOROGNA: *A relaxation theorem and its applications to the equilibrium of gases*. Arch.Rational Mech.Anal., **77** (1981), 359-386.

[110] B.DACOROGNA: *Weak Continuity and Weak Lower Semicontinuity of Non-linear Functionals*. Lecture Notes in Math. **922**, Springer-Verlag, Berlin (1982).

[111] B.DACOROGNA: *Quasi convexité et semicontinuité inférieure faible des fonctionnelles non linéaires*. Ann.Scuola Norm.Sup.Pisa Cl.Sci., **9** (1982), 627-644.

[112] B.DACOROGNA: *Minimal hypersurfaces problems in parametric form with nonconvex integrands*. Indiana Univ.Math.J., **31** (1982), 531-552.

[113] B.DACOROGNA: *Quasiconvexity and relaxation of nonconvex problems in the calculus of variations*. J.Funct.Anal., **46** (1982), 102-118.

[114] B.DACOROGNA: *Remarques sur les notions de policonvexité, quasi-convexité et convexité de rang 1*. J.Math.Pures Appl., **64** (1985), 403-438.

[115] B.DACOROGNA: *Relaxation for some dynamical problems*. Proc.Roy.Soc. Edinburgh, **A-100** (1985), 39-52.

[116] B.DACOROGNA: *Convexity of certain integrals of the calculus of variations*. Proc.Roy.Soc.Edinburgh, **107-A** (1987), 15-26.

[117] B.DACOROGNA: *Direct Methods in the Calculus of Variations*. Book in preparation.

[118] B.DACOROGNA & N.FUSCO: *Semicontinuité des fonctionnelles avec contraintes du type "*$\det \nabla u > 0$*"*. Boll.Un.Mat.Ital., **4-B** (1985), 179-189.

[119] G.DAL MASO: *Integral representation on* $BV(\Omega)$ *of* Γ*-limits of variational integrals*. Manuscripta Math., **30** (1980), 387-413.

[120] G.DAL MASO: *On the integral representation of certain local functionals*. Ricerche Mat., **32** (1983), 85-131.

[121] G.DAL MASO & E.DE GIORGI & L.MODICA: *Weak convergence of measures on spaces of lower semicontinuous functions*. Proceedings of "Integral Functionals in Calculus of Variations", Trieste 1985, Rend.Circ.Mat.Palermo, Suppl. **15** (1987), 59-100.

[122] G.DAL MASO & L.MODICA: *A general theory of variational integrals*. Quaderno della Scuola Normale Superiore "Topics in Functional Analysis 1980-81", Pisa (1982), 149-221.

[123] G.DAL MASO & L.MODICA: *Sulla convergenza dei minimi locali*. Boll.Un. Mat.Ital., **1-A** (1982), 55-61.

[124] G.DAL MASO & L.MODICA: *Integral functionals determined by their minima*. Rend.Sem.Mat.Univ.Padova, **76** (1986), 255-267.

[125] G.DAL MASO & U.MOSCO: *Wiener criteria and energy decay for relaxed Dirichlet problems*. Arch.Rational Mech.Anal., **95** (1986), 345-387.

[126] E.DE GIORGI: *Teoremi di semicontinuità nel calcolo delle variazioni*. Notes of a course held at the Istituto Nazionale di Alta Matematica, Roma (1968-69).

[127] E.DE GIORGI: *Convergence problems for functionals and operators*. Proceedings of "Recent Methods in Nonlinear Analysis", Rome 1978, edited by E.De Giorgi & E.Magenes & U.Mosco, Pitagora, Bologna (1979), 131-188.

[128] E.DE GIORGI: *G-operators and Γ-convergence*. Proceedings of "International Congress of Mathematicians", Warszawa 1983, North-Holland, Amsterdam (1984), 1175-1191.

[129] E.DE GIORGI: *Some semicontinuity and relaxation problems*. Proceedings of "Ennio De Giorgi Colloquium", Paris 1983, edited by P.Kree, Res.Notes in Math. **125**, Pitman, Boston (1985), 1-11.

[130] E.DE GIORGI & L.AMBROSIO & G.BUTTAZZO: *Integral representation and relaxation for functionals defined on measures*. Atti Accad.Naz.Lincei Rend.Cl.Sci.Fis.Mat. Natur., (to appear).

[131] E.DE GIORGI & G.BUTTAZZO: *Limiti generalizzati e loro applicazione alle equazioni differenziali*. Matematiche, **36** (1981), 53-64.

[132] E.DE GIORGI & G.BUTTAZZO & G.DAL MASO: *On the lower semicontinuity of certain integral functionals*. Atti Accad.Naz.Lincei Rend.Cl.Sci.Fis. Mat.Natur., **74** (1983), 274-282.

[133] E.DE GIORGI & G.DAL MASO: *Γ-convergence and calculus of variations.* Proceedings of "Mathematical Theories of Optimization", S.Margherita Ligure 1981, edited by J.P.Cecconi & T.Zolezzi, Lectures Notes in Math. **979**, Springer-Verlag, Berlin (1983), 121-143.

[134] E.DE GIORGI & T.FRANZONI: *Su un tipo di convergenza variazionale.* Atti Accad.Naz.Lincei Rend.Cl.Sci.Fis.Mat.Natur., **58** (1975), 842-850.

[135] E.DE GIORGI & T.FRANZONI: *Su un tipo di convergenza variazionale.* Rend.Sem.Mat.Brescia, **3** (1979), 63-101.

[136] E.DE GIORGI & G.LETTA: *Une notion générale de convergence faible pour des fonctions croissantes d'ensemble.* Ann.Scuola Norm.Sup.Pisa Cl.Sci., **4** (1977), 61-99.

[137] C.J.DE LA VALLEE POUSSIN: *Sur l'intégrale de Lebesgue.* Trans.Amer. Math.Soc., **16** (1915), 435-501.

[138] C.DELLACHERIE & P.MEYER: *Probabilities and Potential.* North-Holland, Amsterdam (1978).

[139] F.DEMENGEL & R.TEMAM: *Convex functions of a measure and applications.* Indiana Univ.Math.J., **33** (1984), 673-709.

[140] M.DOLCHER: *Topologie e strutture di convergenza.* Ann.Scuola Norm.Sup. Pisa Cl.Sci., **14** (1960), 63-92.

[141] S.DOLECKI: *Remarks on semicontinuity.* Bull.Polish Acad.Sci.Math., **25** (1977), 863-867.

[142] L.DREWNOWSKI & W.ORLICZ: *On orthogonally additive functionals.* Bull. Polish Acad.Sci.Math., **16** (1968), 883-888.

[143] L.DREWNOWSKI & W.ORLICZ: *Continuity and representation of orthogonally additive functionals.* Bull.Polish Acad.Sci.Math., **17** (1969), 647-653.

[144] N.DUNFORD & J.T.SCHWARTZ: *Linear Operators.* Interscience Publishers Inc., New York (1957).

[145] G.EISEN: *A counterexample for some lower semicontinuity results.* Math.Z., **162** (1978), 141-144.

[146] G.EISEN: *A selection lemma for sequences of measurable sets, and lower semicontinuity of multiple integrals.* Manuscripta Math., **27** (1979), 73-79.

[147] I.EKELAND: *Nonconvex minimization problems.* Bull.Amer.Math.Soc., **1** (1979), 443-475.

[148] I.EKELAND & R.TEMAM: *Convex Analysis and Variational Problems.* North-Holland, Amsterdam (1976).

[149] S.FANELLI: *Un'estensione di un teorema di semicontinuità inferiore*. Boll. Un.Mat.Ital., **14-B** (1977), 370-382.

[150] H.FEDERER: *Some properties of distributions whose partial derivatives are representable by integration*. Bull.Amer.Math.Soc., **74** (1968), 183-186.

[151] H.FEDERER: *Geometric Measure Theory*. Springer-Verlag, Berlin (1969).

[152] H.FEDERER: *Colloquium lectures on geometric measure theory*. Bull.Amer.Math.Soc., **84** (1978), 291-338.

[153] H.FEDERER & W.ZIEMER: *The Lebesgue set of a function whose distribution derivatives are p-th power summable*. Indiana Univ.Math.J., **22** (1972), 139-158.

[154] F.FERRO: *Integral characterization of functionals defined on spaces of BV functions*. Rend.Sem.Mat.Univ.Padova, **61** (1979), 177-203.

[155] F.FERRO: *Lower semicontinuity, optimization and regularizing extensions of integral functionals*. SIAM J.Control Optim., **19** (1981), 433-444.

[156] F.FERRO: *Lower semicontinuity of integral functionals and applications*. Boll. Un.Mat.Ital., **1-B** (1982), 753-763.

[157] G.FICHERA: *Semicontinuity of multiple integrals in ordinary form*. Arch. Rational Mech.Anal., **17** (1964), 339-352.

[158] A.FOUGERES & A.TRUFFERT: *Δ-integrands and essential infimum, Nemyckii representation of l.s.c. operators on decomposable spaces and Radon-Nikodym-Hiai representation of measure functionals*. Preprint A.V.A.M.A.C. University of Perpignan, Perpignan (1984).

[159] A.FOUGERES & A.TRUFFERT: *Applications des méthodes de représentation intégrale et d'approximation inf-convolutives à l'épi-convergence*. Preprint A.V.A.M.A.C. University of Perpignan, Perpignan (1985).

[160] N.FRIEDMAN & M.KATZ: *Additive functionals of L^p spaces*. Canad.J. Math., **18** (1966), 1264-1271.

[161] N.FUSCO: *Dualità e semicontinuità per integrali del tipo dell'area*. Rend. Accad.Sci.Fis.Mat.Napoli, **46** (1979), 81-90.

[162] N.FUSCO: *Quasiconvessità e semicontinuità per integrali multipli di ordine superiore*. Ricerche Mat., **29** (1980), 307-323.

[163] N.FUSCO & G.MOSCARIELLO: *L^2-lower semicontinuity of functionals of quadratic type*. Ann.Mat.Pura Appl., **129** (1981), 305-326.

[164] A.GAVIOLI: *A lower semicontinuity theorem for the integral of the calculus of variations*. Atti Sem.Mat.Fis.Univ.Modena, **31** (1982), 268-284.

[165] A.GAVIOLI: *Necessary and sufficient conditions for the lower semicontinuity of certain integral functionals*. Ann.Univ.Ferrara, (to appear).

[166] M.GIAQUINTA: *Multiple Integrals in the Calculus of Variations and Nonlinear Elliptic Systems*. Princeton University Press, Princeton (1983).

[167] M.GIAQUINTA & E.GIUSTI: *Q-minima*. Ann.Inst.H.Poincaré Anal.Non Linéaire, **1** (1984), 79-107.

[168] M.GIAQUINTA & G.MODICA & J.SOUCEK: *Functionals with linear growth in the calculus of variations I.II*. Comment.Math.Univ.Carolin., **20** (1979), 143-156 and 157-172.

[169] D.GILBARG & N.S.TRUDINGER: *Elliptic Partial Differential Equations of Second Order*. Springer-Verlag, Berlin (1977).

[170] E.GIUSTI: *Superfici cartesiane di area minima*. Rend.Sem.Mat.Fis.Milano, **40** (1970), 135-153.

[171] E.GIUSTI: *Minimal Surfaces and Functions of Bounded Variation*. Birkhauser-Verlag, Basel (1984).

[172] C.GOFFMAN: *Lower semicontinuity and area functionals I. The non-parametric case*. Rend.Circ.Mat.Palermo, **2** (1953), 203-235.

[173] C.GOFFMAN & J.SERRIN: *Sublinear functions of measures and variational integrals*. Duke Math.J., **31** (1964), 159-178.

[174] H.H.GOLDSTINE: *A History of the Calculus of Variations from the 17th to the 19th Century*. Springer-Verlag, Berlin (1980).

[175] J.HADAMARD: *Sur quelques questions du calcul des variations*. Bull.Soc. Math.France, **33** (1905), 73-80.

[176] T.HADRI: *Fonction convexe d'une mesure*. C.R.Acad.Sci.Paris, **I-301** (1985), 687-690.

[177] M.R.HESTENES: *Sufficient conditions for multiple integral problems in the calculus of variations*. Amer.J.Math., **70** (1948), 239-276.

[178] M.R.HESTENES: *Calculus of Variations and Optimal Control Theory*. Wiley, New York (1966).

[179] M.R.HESTENES & E.J.MC SHANE: *A theorem on quadratic forms and its applications in the calculus of variations*. Trans.Amer.Math.Soc., **47** (1949), 501-512.

[180] F.HIAI: *Representation of additive functionals on vector valued normed Kothe spaces*. Kodai Math.J., **2** (1979), 300-313.

[181] A.D.IOFFE: *An existence theorem for a general Bolza problem.* SIAM J.Control Optim., **14** (1976), 458-466.

[182] A.D.IOFFE: *Sur la semicontinuité des fonctionnelles intégrales.* C.R.Acad.Sci. Paris Sér.A, **284** (1977), 807-809.

[183] A.D.IOFFE: *On lower semicontinuity of integral functionals I.II.* SIAM J.Control Optim., **15** (1977), 521-538 and 991-1000.

[184] A.D.IOFFE & V.M.TIHOMIROV: *Theory of Extremal Problems.* North-Holland, Amsterdam (1979).

[185] W.KARUSH: *A semi-strong minimum for a multiple integral problem in the calculus of variations.* Trans.Amer.Math.Soc., **63** (1948), 439-451.

[186] J.L.KELLEY: *General Topology.* Van Nostrand, Toronto (1955).

[187] D.KINDERLEHRER & G.STAMPACCHIA: *Introduction to variational inequalities and their applications.* Academic Press, New York (1980).

[188] R.V.KOHN & G.STRANG: *Explicit relaxation of a variational problem in optimal design.* Bull.Amer.Math.Soc., **9** (1983), 211-214.

[189] R.V.KOHN & G.STRANG: *Optimal design and relaxation of variational problems. I.II.* Comm.Pure Appl.Math., **39** (1986), 113-137, 139-182, and 353-377.

[190] M.LAVRENTIEV: *Sur quelques problèmes du calcul des variations.* Ann.Mat. Pura Appl., **4** (1926), 7-28.

[191] E.B.LEE & L.MARCUS: *Foundations of Optimal Control Theory.* John Wiley & Sons, London (1968).

[192] J.L.LIONS & E.MAGENES: *Non-homogeneous Boundary Value Problems and Applications. I.II.III.* Springer-Verlag, Berlin (1972) and (1973).

[193] F.C.LIU: *A Luzin type property of Sobolev functions.* Indiana Univ.Math.J., **26** (1977), 645-651.

[194] K.A.LURIE & A.V.CHERKAEV: *Optimal structural design and relaxed controls.* Optimal Control Appl.Methods, **4** (1983), 387-392.

[195] B.MANIA': *Sopra un esempio di Lavrentieff.* Boll.Un.Mat.Ital., **13** (1934), 147-153.

[196] P.MARCELLINI: *Some problems of semicontinuity.* Proceedings of "Recent Methods in Nonlinear Analysis", Rome 1978, edited by E.De Giorgi & E. Magenes & U.Mosco, Pitagora, Bologna (1979), 205- 222.

[197] P.MARCELLINI: *Alcune osservazioni sull'esistenza del minimo di integrali del calcolo delle variazioni senza ipotesi di convessità*. Rend.Mat., **13** (1980), 271-281.

[198] P.MARCELLINI: *Quasiconvex quadratic forms in two dimensions*. Appl. Math.Optim., **11** (1984), 183-189.

[199] P.MARCELLINI: *Approximation of quasiconvex functions, and lower semicontinuity of multiple integrals*. Manuscripta Math., **51** (1985), 1-28.

[200] P.MARCELLINI: *On the definition and the lower semicontinuity of certain quasiconvex integrals*. Ann.Inst.H.Poincaré Anal.Non Linéaire, **3** (1986), 391-409.

[201] P.MARCELLINI & C.SBORDONE: *Relaxation of nonconvex variational problems*. Atti Accad.Naz.Lincei Rend.Cl.Sci.Fis.Mat.Natur., **63** (1977), 341-344.

[202] P.MARCELLINI & C.SBORDONE: *Dualità e perturbazioni di funzionali integrali*. Ricerche Mat., **26** (1977), 383-421.

[203] P.MARCELLINI & C.SBORDONE: *Semicontinuity problems in the calculus of variations*. Nonlinear Anal., **4** (1980), 241-257.

[204] P.MARCELLINI & C.SBORDONE: *On the existence of minima of multiple integrals of the calculus of variations*. J.Math.Pures Appl., **62** (1983), 1-10.

[205] M.MARCUS & V.J.MIZEL: *Absolute continuity on tracks and mappings of Sobolev spaces*. Arch.Rational Mech.Anal., **45** (1972), 294-320.

[206] M.MARCUS & V.J.MIZEL: *Nemyckii operators on Sobolev spaces*. Arch.Rational Mech.Anal., **51** (1973), 347-370.

[207] M.MARCUS & V.J.MIZEL: *Transformation by functions in Sobolev spaces and lower semicontinuity for parametric variational problems*. Bull.Amer.Math. Soc., **79** (1973), 790-795.

[208] M.MARCUS & V.J.MIZEL: *Lower semicontinuity in parametric variational problems, the area formula and related results*. Amer.J.Math., **99** (1975), 579-600.

[209] M.MARCUS & V.J.MIZEL: *Continuity of certain Nemyckii operators on Sobolev spaces and the chain rule*. J.Analyse Math., **28** (1975), 303-334.

[210] M.MARCUS & V.J.MIZEL: *Extension theorems for nonlinear disjointly additive functionals and operators on Lebesgue spaces, with applications*. Bull. Amer.Math.Soc., **82** (1976), 115-117.

[211] M.MARCUS & V.J.MIZEL: *A characterization of nonlinear functionals on $W^{1,p}$ possessing autonomous kernels*. Pacific J.Math., **65** (1976), 135-158.

[212] M.MARCUS & V.J.MIZEL: *Extension theorems of Hahn-Banach type for nonlinear disjointly additive functionals and operators in Lebesgue spaces*. J. Funct.Anal., **24** (1977), 303-335.

[213] M.MARCUS & V.J.MIZEL: *Representation theorems for nonlinear disjointly additive functionals and operators on Sobolev spaces*. Trans.Amer.Math.Soc., **228** (1977), 1-45.

[214] M.MARCUS & V.J.MIZEL: *Every superposition operator mapping one Sobolev space into another is continuous*. J.Funct.Anal., **33** (1979), 217-229.

[215] M.MARCUS & V.J.MIZEL: *A characterization of first order nonlinear partial differential operators on Sobolev spaces*. J.Funct.Anal., **38** (1980), 118-138.

[216] T.MARUYAMA: *Continuity theorem for nonlinear integral functionals and Aumann-Perles' variational problem*. Proc.Japan Acad., **A-62** (1986), 163-165.

[217] E.MASCOLO & R.SCHIANCHI: *Existence theorems for nonconvex problems*. J.Math.Pures Appl., **62** (1983), 349-359.

[218] E.MASCOLO & R.SCHIANCHI: *Nonconvex problems of the calculus of variations*. Nonlinear Anal., **9** (1985), 371-380.

[219] U.MASSARI & M.MIRANDA: *Minimal Surfaces of Codimension One*. Notas de Matematica, North-Holland, Amsterdam (1984).

[220] P.MATTILA: *Lower semicontinuity, existence and regularity theorems for elliptic variational integrals of multiple valued functions*. Trans.Amer.Math.Soc., **280** (1983), 589-610.

[221] V.G.MAZ'YA: *Sobolev spaces*. Springer Verlag, Berlin (1985).

[222] E.J.MC SHANE: *Relaxed controls and variational problems*. SIAM J.Control Optim., **5** (1967), 438-485.

[223] N.G.MEYERS: *Quasiconvexity and lower semicontinuity of multiple variational integrals of any order*. Trans.Amer.Math.Soc., **119** (1965), 125-149.

[224] N.G.MEYERS & J.SERRIN: $H=W$. Proc.Nat.Acad.Sci.U.S.A., **51** (1964), 1055-1056.

[225] E.MICHAEL: *Continuous selections I.II*. Ann.of Math., **63** (1956), 361-382 and **64** (1956), 562-580.

[226] M.MIRANDA: *Distribuzioni aventi derivate misure e insiemi di perimetro localmente finito*. Ann.Scuola Norm.Sup.Pisa Cl.Sci., **18** (1964), 27-56.

[227] M.MIRANDA: *Sul minimo dell'integrale del gradiente di una funzione*. Ann. Scuola Norm.Sup.Pisa Cl.Sci., **19** (1965), 626-665.

[228] M.MIRANDA: *Comportamento delle successioni convergenti di frontiere minimali*. Rend.Sem.Mat.Univ.Padova, **38** (1967), 238-257.

[229] M.MIRANDA: *Dirichlet problem with L^1 data for the non homogeneous minimal surfaces equation*. Indiana Univ.Math.J., **24** (1974), 227-241.

[230] V.J.MIZEL: *Characterization of nonlinear transformations possessing kernels*. Canad.J.Math., **22** (1970), 449-471.

[231] V.J.MIZEL & K.SUNDARESAN: *Representation of vector valued nonlinear functions*. Trans.Amer.Math.Soc., **159** (1971), 111-127.

[232] K.V.MOROZOV: *Lower semicontinuous extension of multidimensional variational problems*. Math.Notes, **38** (1985), 693-699.

[233] C.B.MORREY: *Quasiconvexity and the semicontinuity of multiple integrals*. Pacific J.Math., **2** (1952), 25-53.

[234] C.B.MORREY: *Multiple Integrals in the Calculus of Variations*. Springer-Verlag, Berlin (1966).

[235] U.MOSCO: *Convergence of convex sets and solutions of variational inequalities*. Adv.in Math., **3** (1969), 510-585.

[236] F.MURAT: *Contre-exemples pour divers problèmes où le contrôle intervient dans les coefficients*. Ann.Mat.Pura Appl., **112** (1977), 49-68.

[237] J.NECAS: *Les Méthodes Directes en Théorie des Equations Elliptiques*. Masson, Paris (1967).

[238] C.OLECH: *The characterization of the weak* closure of certain sets of integrable functions*. SIAM J.Control Optim., **122** (1974), 311-318.

[239] C.OLECH: *Existence theory in optimal control problems, the underlying ideas*. Proceedings of "International Conference on Differential Equations", Academic Press, New York (1975), 612-629.

[240] C.OLECH: *Weak lower semicontinuity of integral functionals*. J.Optim.Theory Appl., **19** (1976), 3-16.

[241] C.OLECH: *A characterization of L^1 weak lower semicontinuity of integral functionals*. Bull.Polish Acad.Sci.Math., **25** (1977), 135-142.

[242] P.OPPEZZI: *Convessità dell'integranda in un funzionale del calcolo delle variazioni*. Boll.Un.Mat.Ital., **1-B** (1982), 763-777.

[243] L.PARS: *An Introduction to the Calculus of Variations*. Heinemann, London (1962).

[244] C.Y.PAUC: *La Méthode Métrique en Calcul des Variations*. Hermann, Paris (1941).

[245] B.T.POLJAK: *Semicontinuity of integral functionals and existence theorems on extremal problems*. Math.USSR Sb., **7** (1969), 65-84.

[246] J.P.RAYMOND: *Conditions nécessaires et suffisantes d'existence de solutions en calcul des variations*. Ann.Inst.H.Poincaré Anal.Non Linéaire, **4** (1987), 169-202.

[247] Y.RESHETNIAK: *General theorems on semicontinuity and on convergence with a functional*. Siberian Math.J., **8** (1967), 801-816.

[248] Y.RESHETNIAK: *Mappings with bounded deformation as extremals of Dirichlet type integrals*. Siberian Math.J., **9** (1968), 487-498.

[249] Y.RESHETNIAK: *Weak convergence of completely additive vector functions on a set*. Siberian Math.J., **9** (1968), 1039-1045.

[250] R.T.ROCKAFELLAR: *Integrals which are convex functionals I.II*. Pacific J. Math., **24** (1968), 525-539 and **39** (1971), 439-469.

[251] R.T.ROCKAFELLAR: *Convex Analysis*. Princeton University Press, Princeton (1972).

[252] R.T.ROCKAFELLAR: *Integral functionals, normal integrands and measurable selections*. Proceedings of "Nonlinear Operators and the Calculus of Variations", Bruxelles 1975, edited by J.P.Gossez & E.J. Lami Dozo & J.Mawhin & L.Waelbroeck, Lecture Notes in Math. **543**, Springer-Verlag, Berlin (1976), 157-207.

[253] W.RUDIN: *Real and Complex Analysis*. Mc Graw-Hill, New York (1966).

[254] W.RUDIN: *Functional Analysis*. Mc Graw-Hill, New York (1973).

[255] E.SANCHEZ-PALENCIA: *Non Homogeneous Media and Vibration Theory*. Lecture Notes in Phys. **127**, Springer-Verlag, Berlin (1980).

[256] C.SBORDONE: *Su una caratterizzazione degli operatori differenziali del 2° ordine*. Atti Accad.Naz.Lincei Rend.Cl.Sci.Fis.Mat.Natur., **54** (1974), 365-372.

[257] C.SBORDONE: *Sulla caratterizzazione degli operatori differenziali del 2° ordine di tipo ellittico*. Rend.Accad.Sci.Fis.Mat.Napoli, **41** (1975), 31-45.

[258] C.SBORDONE: *Su alcune applicazioni di un tipo di convergenza variazionale*. Ann.Scuola Norm.Sup.Pisa Cl.Sci., **2** (1975), 617-638.

[259] D.SERRE: *Formes quadratiques et calcul des variations*. J.Math.Pures Appl., **62** (1983), 177-196.

[260] J.SERRIN: *On a fundamental theorem of the calculus of variations*. Acta Math., **102** (1959), 1-22.

[261] J.SERRIN: *A new definition of the integral for non-parametric problems in calculus of variations*. Acta Math., **102** (1959), 23-32.

[262] J.SERRIN: *On the definition and properties of certain variational integrals*. Trans.Amer.Math.Soc., **101** (1961), 139-167.

[263] J.SERRIN & D.E.VARBERG: *A general chain rule for derivatives and the change of variables formula for the Lebesgue integral*. Amer.Math.Monthly, **76** (1969), 514-520.

[264] E.SILVERMAN: *A sufficient condition for the lower semicontinuity of parametric integrals*. Trans.Amer.Math.Soc., **167** (1972), 465-469.

[265] E.SILVERMAN: *Strong quasi-convexity*. Pacific J.Math., **46** (1973), 549-554.

[266] E.SILVERMAN: *A necessary condition in the calculus of variations*. Proc. Amer.Math.Soc., **37** (1973), 462-464.

[267] E.SILVERMAN: *Lower semicontinuity of parametric integrals*. Trans.Amer. Math.Soc., **175** (1973), 499-508.

[268] E.SILVERMAN: *Basic calculus of variations*. Pacific J.Math., **104** (1983), 471-482.

[269] S.SPAGNOLO: *Una caratterizzazione degli operatori differenziali autoaggiunti del 2° ordine a coefficienti misurabili e limitati*. Rend.Sem.Mat.Univ.Padova, **38** (1967), 238-257.

[270] I.V.SRAGIN: *Abstract Nemyckii operators are locally defined operators*. Soviet Math.Dokl., **17** (1976), 354-357.

[271] A.W.STODDART: *Semicontinuity of integrals*. Trans.Amer.Math.Soc., **122** (1966), 120-135.

[272] G.STRANG & R.TEMAM: *Functions of bounded deformations*. Arch.Rational Mech.Anal., **75** (1980), 7-21.

[273] R.TEMAM: *Solutions généralisées des certaines équations du type hypersurfaces minima*. Arch.Rational Mech.Anal., **44** (1971), 248-264.

[274] R.TEMAM: *A characterization of quasi-convex functions*. Appl.Math.Optim., **8** (1982), 287-291.

[275] R.TEMAM: *Approximation de fonctions convexes sur un espace de mesures et applications*. Canad.Math.Bull., **25** (1982), 392-413.

[276] R.TEMAM & G.STRANG: *Duality and relaxation in the variational problems of plasticity*. J.Méc.Théor.Appl., **19** (1980), 493-527.

[277] L.TONELLI: *Sur une méthòde directe du calcul des variations*. Rend.Circ.Mat. Palermo, **39** (1915), 223-264.

[278] L.TONELLI: *Fondamenti di Calcolo delle Variazioni. I.II*. Zanichelli, Bologna (1921) and (1923).

[279] L.TONELLI: *Sugli integrali del calcolo delle variazioni in forma ordinaria*. Ann.Scuola Norm.Sup.Pisa Cl.Sci., **2** (1934), 401-450.

[280] L.TONELLI: *Opere Scelte. I.II.III.IV*. Cremonese, Roma (1960),(1961), (1962),(1963).

[281] TRAN CAO NGUYEN: *A characterization of some weak semicontinuity of integral functionals*. Studia Math., **66** (1979), 81-92.

[282] M.M.VAINBERG: *Variational Methods for the Study of Nonlinear Operators*. Holden-Day, San Francisco (1964).

[283] M.VALADIER: *Closedness in the weak topology of the dual pair L^1, C*. J. Math.Anal.Appl., **69** (1979), 17-34.

[284] M.VALADIER: *Régularisation s.c.i., relaxation et theorèmes bang-bang*. C.R. Acad.Sci.Paris, **I-293** (1981), 115-116.

[285] G.N.VASILENKO: *Weakly continuous functionals of the calculus of variations in the spaces $W^{1,p}(U,\mathbf{R}^m)$*. Soviet Math.Dokl., **32** (1985), 706-709.

[286] M.VALADIER: *Fonctions et opérateurs sur les mesures*. C.R.Acad.Sci.Paris, **I-304** (1987), 135-137.

[287] J.WARGA: *Relaxed variational problems*. J.Math.Anal.Appl., **4** (1962), 111-128.

[288] J.WARGA: *Necessary conditions for minimum in relaxed variational problems*. J.Math.Anal.Appl., **4** (1962), 129-145.

[289] J.WARGA: *Optimal Control of Differential and Functional Equations*. Academic Press, New York (1972).

[290] W.A.WOYCZYNSKI: *Additive functionals on Orlicz spaces*. Colloq.Math., **19** (1968), 319-326.

[291] K.YOSIDA: *Functional Analysis*. Springer-Verlag, Berlin (1980).

[292] L.C.YOUNG: *Generalized surfaces in the calculus of variations. I.II*. Ann.of Math., **43** (1942), 84-103 and 530-544.

[293] L.C.YOUNG: *Lectures on the Calculus of Variations*. W.B.Saunders, Philadelphia (1969).

[294] T.ZOLEZZI: *On equiwellset minimum problems*. Appl.Math.Optim., **4** (1978), 209-223.

Index